Introduction to
DIGITAL DATA COMMUNICATIONS

Introduction to
DIGITAL DATA COMMUNICATIONS

David H. Stein

Delmar Publishers Inc.

0-8273-2436-7

90000

9 780827 324367

to Barbara, Steve, and Gary

For information, address Delmar Publishers Inc.
2 Computer Drive West, Box 15-015
Albany, New York 12212-5015

COPYRIGHT © 1985
BY DELMAR PUBLISHERS INC.

Printed in the United States of America
Published simultaneously in Canada
by Nelson Canada,
A division of The Thomson Corporation

10 9 8 7 6

Library of Congress Cataloging in Publication Data

Stein, David H.
   Introduction to digital data communications.

   Includes index.
   1. Data transmission systems.   2. Digital
communications.   I. Title.
TK5105.S74   1985   384   84-20072
ISBN 0-8273-2436-7

Contents

Preface

This book is based on a data communications systems course taught to graduating students in an electronic technology A.A.S. program at Heald College. Its writing was motivated by the difficulty in finding a text suitable for junior college students among existing publications in this field. The available books appear to fall into two categories. Some are intended for advanced undergraduate or graduate level university students and are too theoretical and mathematically complex. Others seem to be aimed at management and engineering personnel currently working in the data processing industry. These were also considered unsuitable for several reasons:

- They provide few, if any, questions, problems, and computational exercises to reinforce the material in the text.
- There is too much emphasis on quasi-technical matters such as F.C.C. regulations and tariffs.
- They take much for granted in terms of the reader's background knowledge of computer systems.

Therefore, the development of this book was guided by the following principles:

- Digital data communications cannot be understood in any meaningful sense if it is taken out of context. The book therefore digresses into a few areas of computer systems theory which are not usually considered to fall within the realm of "communications."
- New concepts should always be reinforced by concrete example. There is an emphasis on such things as: timing relationships on interface circuits; the appearance of signals on oscilloscope traces; event sequences for protocols; computation of data rates and time required to execute various communications functions.
- This book is not intended to be, and could not possibly be, a survey of all current communications techniques and equipment. I believe it is preferable to select one design approach that illustrates basic principles and to cover it in depth, rather than to superficially study numerous approaches.
- At an introductory level, excessive detail is frequently a source of confusion rather than information. The material therefore attempts to present suffi-

cient detail to illustrate underlying principles, while avoiding both oversimplification and sensory overload.
- For a technical subject, there is no substitute for clear, precise drawings and tables.

Data communications, like many other technologies, is a layered succession of structures, each providing a foundation for those that rest above. I believe that the upper levels cannot be fully appreciated without a clear picture of what lies underneath. The book therefore follows an ascending path. The starting point, communications-oriented, digital integrated circuits, is one at which an electronic technology student can feel comfortable.

After a brief introduction to the nature of communications media and their data formats, a fairly detailed picture of the operation of UARTs is presented. An asynchronous interface for a small computer is built around the UART. This provides a vehicle for an introduction to the principles of communications software. In all cases, general concepts are related to the operation of commercially available devices. The next step "outward" is the interface to the digital side of a modem and an explanation of procedures used to control a communications medium. This naturally leads to a discussion of modulation techniques and the nature of the telephone system as a carrier of digital information.

With this foundation, the text proceeds to more advanced subjects, such as error detection, protocols, and networks. The final chapters introduce a number of "system-level" concepts which are usually not emphasized in electronic technology programs. Recurring themes in the text are

- design constraints imposed on communications equipment by the physical characteristics of the medium;
- the need for standardization and compatibility;
- the relationship of communications equipment and procedures to the functions computers perform for their "end users."

In terms of mathematical background, the text presumes the student is familiar with

- basic algebra
- logarithms
- the functions of basic digital logic circuits
- binary and hexadecimal number systems

The last two subjects are briefly reviewed in the appendixes. Rather than the conventional method of representing algebraic symbols by single subscripted characters, I have chosen to use a notation resembling programming languages. Since this notation allows representation of symbols by multicharacter strings, I believe it substantially improves readability. In this notation a symbol is a string of alphanumeric characters with the possible inclusion of "underscore" as a readability separator (for example, BAUD_RATE). Multiplication is indicated by *. All other arithmetic operators are as usual.

When this course was taught at Heald, it was accompanied by laboratory work, which was an excellent teaching aid. Experiments were based on IBM Personal Computers equipped with Asynchronous Communication Adapters. Use of the Advanced BASIC interpreter allowed complete control of communications parameters without becoming involved with the intricacies of assembly language. Experimental work is not included in this book because it is so highly dependent on the availability of specific hardware. It nevertheless should be strongly considered as a supplement to any course based on this material.

Over the years numerous people have influenced my thinking and shared ideas that found their way into this book. I would like to thank the administration of Heald College, Santa Clara, particularly Dean Edward Tuba and Director Robert Meyers, for giving me the opportunity to teach this course and the freedom to do it as I saw fit. I would also like to thank W. Guy Rowe, who helped establish this course at Heald and who shared the joys and frustrations of teaching it for the first time. To my students I express sincere thanks for instant feedback concerning the presentations which worked and those which did not. I would like to thank the reviewers for their help in developing this text. These reviewers include Thomas J. Bingham, Jr. of St. Louis Community College; Frederick F. Driscoll of Wentworth Institute of Technology; Arthur L. Foston of California State University; Roy H. Kratzer of Foothill College; Tad Kubic of West Valley College; John K. Morrison of Heald College; and Paul H. Young of Arizona State University. Finally, I would like to thank Delmar Publishers and Kip Sears for their faith in this book and its author.

D. Stein
Sunnyvale, CA   1984

chapter one

INTRODUCTION

1·1   OVERVIEW

The most basic definition of communications in general is the transmission of messages from a *sender* to a *receiver* over a *medium*. We can all think of numerous examples. They range over a spectrum of cost and complexity that begins with "tin cans and string" and ends with geosynchronous communications satellites. We will deal with the subdivision of this body of knowledge that involves conversations between computers and similar devices.

Exactly what is data communications? It is an industry, a branch of technology, and most of all, a group of concepts shared by a community of people. In the following chapters we explore these ideas and learn the specialized vocabulary which describes them. It is an important and interesting subject for a number of reasons. The proliferation of personal computers and the development of nationwide public data networks have made communications technology (once limited to the data processing departments of large corporations) available to millions of people. Automated bank teller machines and centralized data bases have already had an impact on the American lifestyle, and this impact will continue to grow for the next few decades. In American industry, the demand for employees with knowledge of communications systems is currently very high and is likely to remain so. These people will be influential policymakers. If we are to make sensible evaluations of the benefits and dangers inherent in this technology, we must first understand what it can and cannot do.

1·2   THE IMPACT OF DISTANCE

The most concise description of the subject matter in this book is the following:

Data communications is a branch of technology concerned with transmission of messages between digital electronic devices separated by a distance so great that conventional digital circuit techniques will not work reliably.

By "conventional digital circuit techniques," we mean

- voltage and current levels that are characteristic of TTL integrated circuits (see Appendix A);
- connections implemented by means of individual wires or printed circuit traces rather than twisted pairs, coaxial cable, or some other medium.

There are two critical points in the previous definition. First, we are discussing *digital devices,* that is, equipment that processes data which is encoded into binary numbers, "ones" (1s) and "zeros" (0s). Second, the element of *distance* is present. In order to understand the impact of distance, let us briefly review digital data transmission when distance is small—for example, inside a personal computer (PC). Figure 1–1 illustrates a typical design which allows a *microprocessor* (the instruction execution circuit) to write (store) data into a memory chip (RAM). In this situation, blocks of 8 bits (bytes) are transferred simultaneously via the wide arrow marked DATA. An individual wire is dedicated to the transmission of each bit. The DATA lines alone are not sufficient to make this transfer work. Control and timing signals are also necessary. These signals (STROBE, ACKNOWLEDGE, and CLOCK) operate as follows:

1. The microprocessor will assert [1] the STROBE line to tell the RAM that *data* is ready and the storage cycle must be initiated.
2. RAM will subsequently assert ACKNOWLEDGE to indicate that storage is complete and the microprocessor may prepare for the next data transfer.
3. The CLOCK is a continuous stream of regularly spaced pulses (a square wave) which is used to synchronize the microprocessor, RAM, and other circuits within the PC. Since all data transfers will occur at the instant the CLOCK "ticks," circuit designers can guarantee that all necessary preparation is complete before the actual transfer takes place.

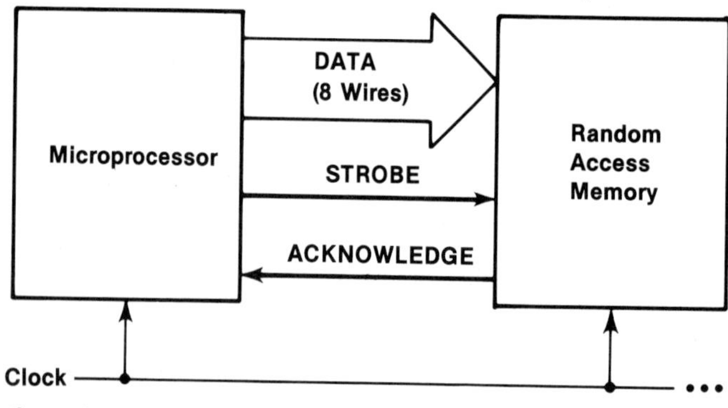

FIGURE 1–1   Short Distance Data Transmission

---

1. In the language of digital systems engineering, a signal is *asserted* when it is turned *on* and *negated* when it is turned *off.*

This is a perfectly reliable communications scheme within the confines of the PC. Suppose we now want to accomplish the same data transfer over a distance of 3000 km. An obvious choice for the communications medium is a telephone line. It is, in general, the only two-way, long-distance communications medium that is already connected to most homes and offices. If you had no prior knowledge about data communications, you might be tempted to reproduce the scheme just described, substituting telephone lines for short wires. We all know that this is not a practical solution. In order to understand exactly why, let us contrast the properties of a short wire with those of a telephone circuit (see Table 1–1).

Bandwidth is the range of signal frequencies that can be passed without attenuation over a transmission medium. The short connections inside the PC, because they approximate ideal conductors, have a wide bandwidth with a lower limit of 0 Hz, that is, direct current. Telephone circuits, on the other hand, were designed to carry signals representing the human voice. Our voices contain only a limited range of frequencies, and standard telephone equipment will pass only a part of this spectrum (300–3300 Hz). This is sufficient to allow a listener to recognize the speaker, but it is hardly high fidelity. Typical signals generated by TTL circuits inside digital equipment contain a much wider spectrum of frequencies than the telephone system can carry. In fact, the faster a digital signal can switch states, the wider the range of frequencies it will contain and the smaller the fraction of its energy that will be transmitted without attenuation by the phone line.

Digital signals travel along wires at speeds approaching the speed of light. Within a PC, the time for a pulse to propagate down the length of a wire (nanoseconds) is insignificant compared to the time required for integrated circuits to perform their functions. Delays on a long-distance telephone circuit, however, are much longer (milliseconds). Many cross-country and international connections now involve satellite links. In this case, delays will approach a half-second. The combination of restricted bandwidth and delay forces us to slow down data transmission rates by at least a factor of 100 when we go from the internal wire to the phone line.

When a TTL circuit transmits a 1 down a few centimeters of wire, it is highly unlikely that the receiving circuit will misinterpret this signal as a 0. Short wires

TABLE 1–1 Properties of Communications Media

PROPERTY	WIRE INSIDE PC	TELEPHONE LINE
Bandwidth (Hz)	$0-10^7$	300–3300
Delay	no	yes
Max Data Rate (bits/s)	$> 10^6$	$< 10^4$
Error Probability	0	moderate
Waveform Distortion	no	yes
Cost Per Bit	low	high

with low impedance make poor antennas for pickup of spurious signals. We can therefore say that the probability of a transmission error within our PC is virtually zero. During the course of telephone conversations, however, we have all heard various clicks, hisses, and, occasionally, the faint echo of someone else's conversation. While these effects are only minor annoyances to human listeners, they can easily corrupt data transmissions. Fortunately, these events are infrequent. The telephone company states that we can expect no more than one bit in 100 000 to be garbled on typical circuits. Nevertheless, we cannot allow critical data (figures representing large sums of money, for example) to be lost: data transmission equipment must provide mechanisms for detecting and retransmitting bad data. In later chapters we shall see that these error recovery procedures can cause a severe degradation in the rate at which data can be sent over a communications line. If error rates were significantly higher than those actually experienced, telephone lines would be useless for data communications.

The problem of waveform distortion is closely related to the narrow bandwidth of the phone line. Typical TTL waveforms contain many frequency components above 1 MHz, as well as a DC component. "Filtering" of the high and low frequency components by the phone line results in distortions of the type illustrated in Figure 1–2. The oscillations (called "ringing") are caused by the elimination of high frequency components, and the "droop" of the pulse tops is caused by the elimination of the DC component. We must therefore convert conventional digital waveforms into signals that fit within the 300–3300 Hz limits of the phone line. Even within these limits we find variations in attenuation and phase shifts, both as a function of frequency, and from one connection to the next. It is possible to compensate for these latter problems, but they make life more difficult for designers of communications equipment.

The bottom line of our comparison table is cost. The internal wire is cheaply purchased when the PC is built, and it carries no continuing cost. In

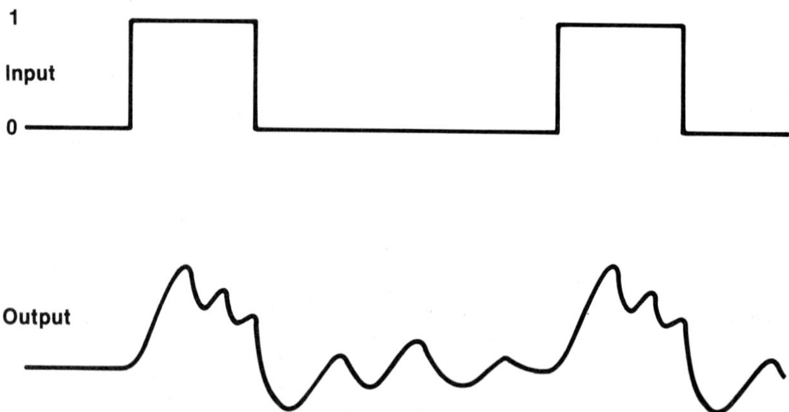

FIGURE 1–2 Waveform Distortion by Bandwidth Limitation

the case of the telephone line, we must pay for every minute it is used. Telephone lines are therefore used sparingly. The block diagram in Figure 1–1 shows a great deal of parallelism. Logic circuits inside the microprocessor and RAM can frequently be simplified due to this separation of data and control signals. Telephone links that carry data usually employ only one circuit (two at most) for two-way communications. Data must be transmitted *serially* (one bit at a time), and we must devise techniques which allow a single circuit to carry data, timing, and control signals simultaneously.

All of the preceding material seems to indicate that data communications is at best a very difficult problem. However, the ability to connect one's terminal or PC to a similar device anywhere in the world is a highly desirable goal, and much ingenuity has been applied to these problems. The following chapters will show how successful this effort has been.

1·3 HARDWARE ELEMENTS FOR COMMUNICATIONS

At this point we briefly introduce the major components of communications systems. All of these subjects will be covered in depth later.

Host Computer

A *host computer* is a medium or large computer system capable of "simultaneously" running programs initiated by several users. In actuality only one

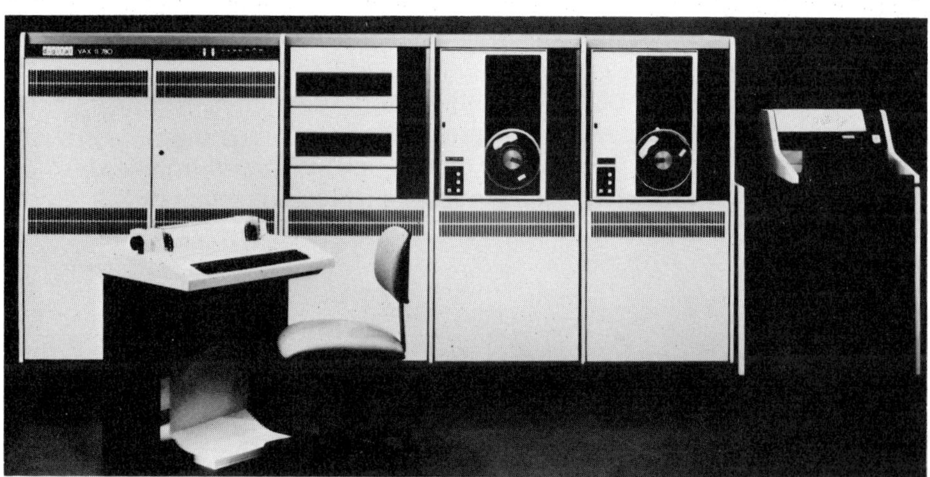

A Host Computer Capable of Supporting Numerous Communications Lines
Photo courtesy of Digital Equipment Corporation

program runs at a time, but the processor switches so rapidly from one program to another that the activity appears simultaneous to the users. The technique is called *time-sharing*. Users are typically connected to the host computer by means of communications lines. The line connectors are called *ports*; a large host computer may have 100 or more ports.

Terminals

A *terminal*, physically separated from the host computer, is a device used for bidirectional data transfer to a host-computer port. It resides at the other end of the communications line that emanates from the host-computer port connector.

The terminal is the place where information from the analog "outside world" is converted to digital format for transmission to the host computer (and vice versa). The source of this external information may be a mechanical device, such as a pressure gauge, but, in the most commonly accepted definition, the ultimate source and recipient of the terminal's data is a human operator. Once again, there are numerous devices which fit into this category. These include supermarket checkout counters; magnetic badge readers that provide security at factory entrances; the familiar general purpose terminal comprised of a keyboard and a cathode ray tube (CRT) display. In the latter case, the analog-to-digital conversion takes place at the instant a finger touches a key.

Modems

In the previous section we noted that digital signals typical of host computers and terminals required conversion to a form that could be carried on a communications medium, such as a telephone line. This conversion process is called *modulation*, and the inverse process (communications medium back to digital) is called *demodulation*. A device that performs these functions is called a *modem*. The *mo* part of the name coming from *mo*dulation; the *dem*, from *dem*odulation. Figure 1–3 illustrates a host computer and a typical collection of terminals and modems.

Networks

In Figure 1–4 we have a collection of *network users* (host computers and terminals) connected to a nebulous thing called a *network*. The connections are communications lines called *access links*, at the ends of which, as you might expect, modems are frequently found.

The purpose of the network is to provide each user with the ability to establish communications with any other user. The most obvious and familiar ex-

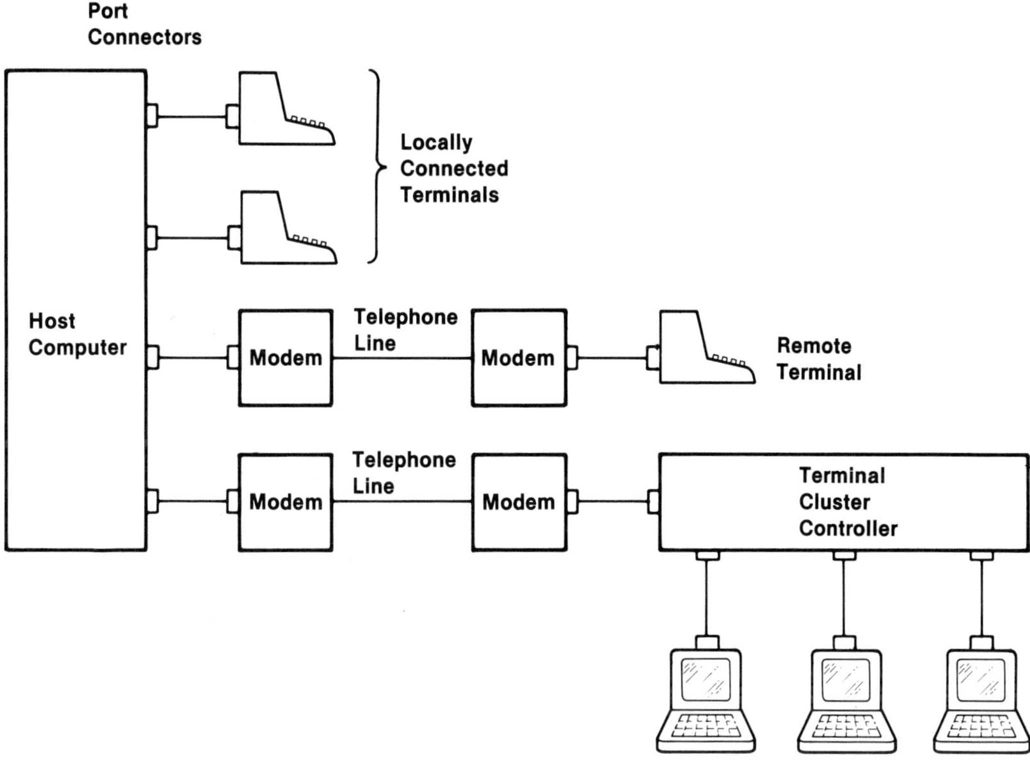

FIGURE 1–3   A Host Computer and Associates

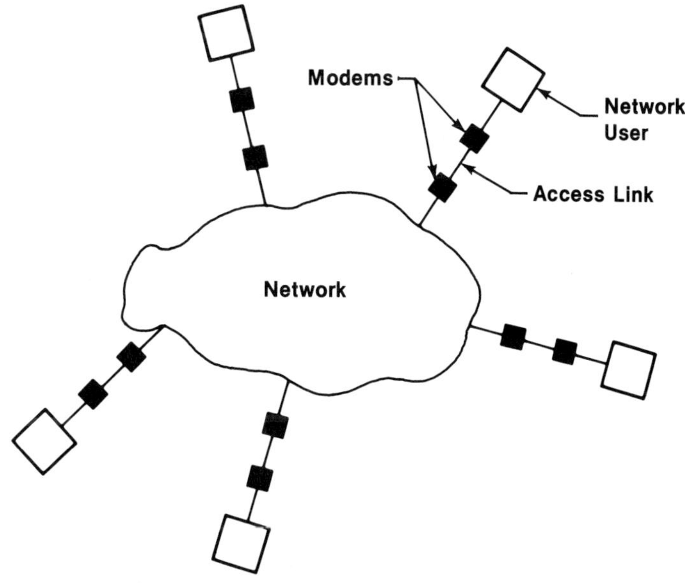

FIGURE 1–4   General Picture of a Network

ample is the telephone system that covers the world. In this book we will distinguish between the two kinds of networks illustrated in Figure 1–5.

The *broadcast network* (Figure 1–5A) is essentially nothing more than a shared communications medium. All users have some sort of "antenna," and any transmission is heard by all other devices on the network. Since messages are usually intended for only one recipient, each transmission must carry an *address* to identify the destination. All other users are obliged to ignore the message. We must also establish a set of rules (a *protocol*) to prevent the situation where several devices are transmitting at the same time. It is critical that all devices on the network obey a common protocol. A communications satellite orbiting above a fixed point on the earth is a good example of this type of network. A transmission from any earth station is received by the satellite and retransmitted in the direction of earth. All earth stations with antennas pointed at the satellite will receive the retransmission.

The *connected network* (Figure 1–5B) is a collection of devices called *nodes*, which are linked by communications circuits called *trunks*. The nodes can establish and maintain a communications path (*connection*) between any pair of users. Many of these connections may be active simultaneously. From the user's point of view, operation on this type of network is much simpler than on a broadcast network. The destination address must only be specified once (at the time a new connection is established), and a user never hears a message destined for someone else. The implementation of the nodes is, of course, a complex and expensive task. There are many possible ways to configure the nodes and interconnecting links. Heavy message traffic may require multiple

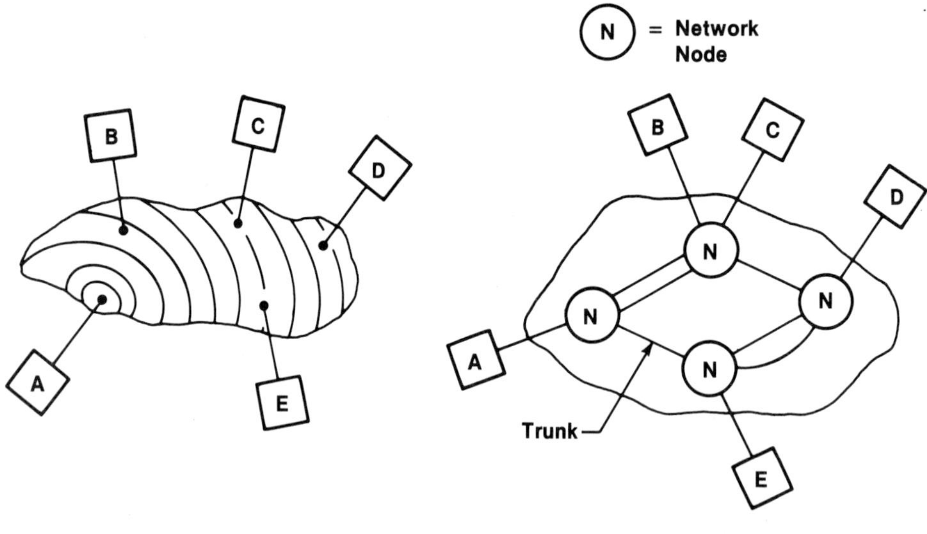

(A) Broadcast Network   (B) Connected Network

FIGURE 1–5   General Network Classes

links between some pairs of nodes. Note that messages between a particular pair of users (for example, between A and D in Figure 1–5B) may have to cross several nodes.

1·4 CIRCUIT TERMINOLOGY

This section presents some commonly encountered terms relating to communications channels (circuits) and the ways they are used. These terms will appear repeatedly in the remainder of the book.

Simplex/Half-duplex/Full-duplex

A channel which transmits data in only one direction is called *simplex*. If transmission is possible in either direction, but only one direction is permitted to be active at any time, the channel is called *half-duplex*. If it is possible to send data in both directions simultaneously, the channel is termed *full-duplex*. In many cases, a circuit may have full-duplex capabilities, but it is used in a half-duplex manner because the equipment attached to it is not capable of simultaneously sending and receiving.

Switched/Dedicated

An ordinary telephone call is an example of the use of a *switched* (or *dial-up*) circuit. The calling party specifies a number, and the phone company's equipment establishes a connection between the two parties. For the entire duration of the call some of the phone company's facilities (wires, microwave links, etc.) are assigned to this connection. When the parties hang up, these facilities are released and available for use by another call. The caller is billed according to the length of time these facilities were used.

At a customer's request, the phone company can also establish a connection between two points and leave it permanently connected. This is called *dedicated* (or *leased*) line, and it is frequently used to link the branch offices of large corporations. The customer pays a fixed monthly rate which is lower than the total cost of a large number of individual calls between the same points. Dedicated circuits can be used for voice or data and offer the advantage of being continuously available. (There are times when it may not be possible to get a switched connection between two busy cities because all links are already in use.)

How is it possible that a dedicated circuit can be less expensive than any number of switched calls? The dedicated circuit uses the phone company's facilities 100% of the time and even extreme usage of switched calls will not begin to approach 100%. The answer lies in the equipment that must be used only at the start of a switched call (devices that interpret dialed digits, for example).

The customer must pay its share for any phone company equipment that is used, and this setup equipment is one of the reasons that the first few minutes of a phone call are the most expensive. Dedicated circuits never require the services of call setup equipment, and this results in cost savings.

When a connection is dedicated, it is also possible for the phone company to reduce its noise level and to control the variations of attenuation and phase shift that cause waveform distortion. This is called *line conditioning,* and it adds to the cost of the dedicated circuit. Conditioning can reduce the error probability and significantly increase the maximum data transfer rate (*throughput*) on the link.

Point-to-Point/Multipoint

Figure 1–6 illustrates these concepts. A point-to-point circuit connects only two devices and may be either switched or dedicated. The *multipoint* (sometimes called *multidrop*) circuit connects several devices which may be widely separated. Multipoint connections are installed by the phone company at a customer's request, and they must be dedicated because of the special devices that are installed at the tap points.

The digital devices attached to a multipoint circuit are called *drops,* and a protocol must be established to prevent interference due to simultaneous transmissions.

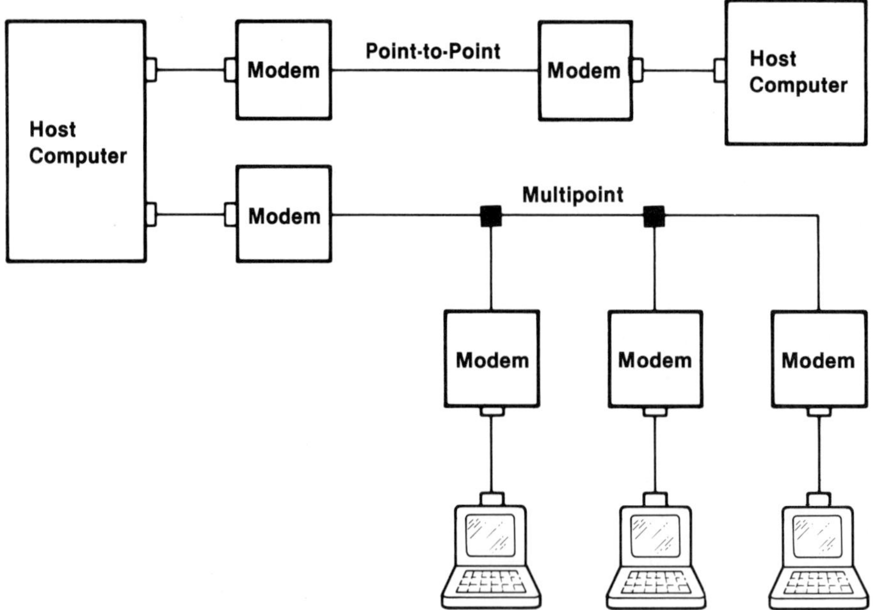

FIGURE 1–6   Point-to-Point and Multipoint Circuit Examples

We have now acquired enough vocabulary to begin examining the details of data communications equipment. If you are unfamiliar with the binary and hexadecimal number systems or would like to review this subject, please read Appendix B before proceeding.

REVIEW QUESTIONS

1. Every new technology brings benefits and dangers to society. What are the possible benefits and dangers of the use of high-speed, nationwide data networks by
   a. police departments
   b. personal computer owners
   c. Internal Revenue Service
   d. public libraries

2. Classify the following communications systems into broadcast or connected categories:
   a. the postal system
   b. a theatrical production
   c. interstate trucking companies
   d. a session of the U.S. Senate
   e. messages sent through pneumatic tubes

3. Classify each of the following broadcast systems as simplex, half-duplex, or full-duplex:
   a. a classroom
   b. a TV station
   c. CB radios
   d. an auction

   In the half-duplex cases describe and contrast the rules which specify when a user is allowed to "transmit" (speak).

4. In recent years, people have proposed using cable-TV systems as two-way media, allowing subscribers to attach terminals or personal computers to the cable for the purpose of transmitting to each other or to host computers. Discuss some of the technical problems that might be involved in setting up such a system.

chapter two

DATA FORMATS

2·1 ASYNCHRONOUS AND SYNCHRONOUS DATA

The information on communications lines is transmitted in short serial bursts of bits called *characters*. A character has a fixed size. Since typical communications equipment will never send a fractional part of a character, they may be considered the indivisible atoms of data communications.

Depending on the equipment in use, a character may be anywhere from 5 bits to 8 bits long. A character is thus a small binary number which is assigned a meaning. The meaning may be *alphanumeric* (a letter, number, or punctuation mark) or a *control function* (for example, a command instructing a printer to advance to the next sheet of paper). A collection of these binary numbers and their associated meanings is called a *code*. There are many codes in use; in later chapters we will see some specific examples. A piece of data communications equipment is usually designed to work in one particular code. Remember that the assignment of meanings to binary character values is a totally arbitrary decision, and the users at both ends of a communications link must agree on the code they are going to use before any meaningful data can be exchanged. This may sound trivial, but a large number of communications problems are caused by a misunderstanding of the exact specifications of the data being sent.

The terms *asynchronous* and *synchronous* are concerned with the time relationships between characters on the line. *Asynchronous* literally means "not synchronized." In an asynchronous ("async" to insiders in the communications business) format, a character can start at any time. After transmission of the character is complete, the line goes into an idle state. There is no upper limit on the length of this idle time. The asynchronous format is therefore well suited to the transmission of data generated by someone typing at a terminal. Each time a key is struck an asynchronous character is transmitted.

In the *synchronous* format, the idle time between characters is eliminated. The first bit of a character comes immediately after the last bit of the previous one. This means that the line is always sending something, and if there is an interval during which the device controlling the line has nothing in particular to say, this time must be filled with a nonalphanumeric character that has been specifically designated for this purpose. There is much more to the issue of synchronous transmission. For example, how can receiving equipment determine where one character ends and the next begins? After all, the only thing we are receiving is a long continuous string of 1s and 0s. We will return to this question in later chapters. The remainder of this chapter will concentrate on the asynchronous format.

2·2 THE SHAPE OF ASYNCHRONOUS DATA

Figure 2–1 is a generalized picture of an asynchronous character as it would appear on an oscilloscope. That is, time advances as we move to the right on the diagram. The first thing we notice is that the two conditions of the binary communications line are called MARK and SPACE rather than 1 and 0. MARK corresponds to 1 and SPACE corresponds to 0. Data communications technology is actually older than electronic computer technology, and some of the early terminology, such as MARK and SPACE, has been carried over.

During the idle time between characters, the line remains in the marking condition. By definition, the transmission of an asynchronous character can be initiated whenever the line is idle. To make this system operable, the receiver must be given some kind of warning to alert it to the fact that data is on the way. This warning is the *start bit,* and it is an example of a way in which a control signal can be carried on the same wire as the data. At the beginning of the start bit, the line goes from MARK to SPACE and remains in the spacing condition for a fixed length of time T (called the *bit time*).

Immediately following the start bit, the data bits are transmitted. They are the actual information content of the character. The least significant bit (LSB) is

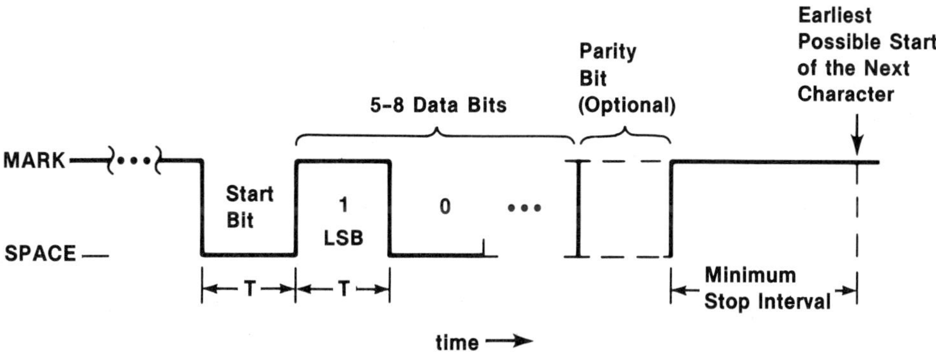

FIGURE 2–1 Generalized Picture of an Asynchronous Character

always transmitted first. The duration of each data bit is T, exactly the same as the duration of the start bit. In fact, the bit time (T) is one of the basic parameters of a communications line. We can compute the line's data rate (in bits per second) as

$$B\_P\_S = \frac{1}{T}$$

For example, if the bit time is 20 ms, we have

$$B\_P\_S = \frac{1}{0.020} = 50 \text{ bits/s}$$

B_P_S is frequently referred to as the line's *baud rate*. Actually, there is a subtle distinction between bits per second and baud rate which we will learn in Chapter 9. As long as we are dealing with a binary (two-state) circuit, the terms can be used interchangeably. When we study the nature of data signals on phone lines, we will find that signal values may have more than two possible states. In these cases there will be a difference in value between B_P_S and baud rate. However, the interface circuits to terminals and host computer ports are invariably binary, and it is common in industry to make no distinction between the two terms. Until we reach Chapter 9 we will follow this practice.

Also note, in Figure 2–1, that the bit values merge. If a sequence of bits of the same polarity is transmitted, there will be no observable event to identify the transition points between bits. It is the time T that determines where one bit ends and the next starts.

We said that, depending on equipment design, there may be 5–8 data bits in a character.   Some equipment is also designed to insert a single bit, called *parity*, immediately after the last data bit in order to provide the receiver with a simple means of checking for errors on the channel. When the character is formed for transmission, the value of the parity bit is selected according to one of the following rules:

Odd parity
Count the number of data bits that are in the MARK (1) state. Select the value of the parity bit such that there is an odd number of 1s in the combination of the data field plus the parity bit (do not include the start bit).

Even parity
Similar to the previous procedure, but the number of 1s in data plus parity is forced to be even rather than odd.

Once again, both transmitter and receiver must agree on which parity rule to follow (if any). Suppose, for example, the odd rule is selected. The receiver will count the number of 1s in each incoming character; if the total is odd, it will assume there were no errors (errors appear as bit values that are changed during transmission). If the count turns out to be even, the receiver will know that a bit has been corrupted. Of course, it is possible that the parity bit itself will be in error, and we will then defame a perfectly good character. This, however, is infrequent and a small price to pay for the ability to detect errors. Exactly what the equipment will do when an error is detected is yet another question that must be decided and agreed upon by the designers at both ends of the communications link. In most cases, the receiver will request that the bad message be retransmitted.

Returning to Figure 2–1, we see that, following the data/parity field, the line returns to the marking state. This is the beginning of the *stop interval* (sometimes erroneously called the *stop bit*). This is really the start of the idle interval between asynchronous characters. To guarantee that there will be a MARK to SPACE transition at the beginning of the start bit for the next character, we must enforce a minimum length for the stop interval. This minimum stop interval is also a variable feature of equipment design and is specified as a multiple of the bit time T. Typical values are 1, 2, and, occasionally, 1.5 times T. Since the format is asynchronous, the stop interval has no maximum length.

2·3 EXAMPLES OF ASYNCHRONOUS CHARACTER WAVEFORMS

To clarify all this, let us look at some typical waveforms produced by the transmission of asynchronous characters (Figure 2–2). Since the binary value of the character is a number somewhere between 5 and 8 bits, it can be represented in two hexadecimal digits. In the first example we have 6 data bits. We therefore expand the hexadecimal value into 8 bits and discard the two most significant bits to get our data pattern. We can now draw the waveform beginning with the MARK-to-SPACE transition at the leading edge of the start bit. We then draw the data bits, remembering that the least significant bit is transmitted first and therefore appears at the left side of the waveform. Bits of similar polarity will merge smoothly; the tick marks at the transition points in Figure 2–2 are shown only to clarify the drawing. Since the first example specified no parity, the stop interval begins immediately after the most significant data bit.

The second example is similar, except that it calls for 7 data bits, and we therefore discard only the most significant bit from the expansion of the hexadecimal value. The example also specifies odd parity. Since the data pattern includes three 1s, the count is already odd, and the parity bit therefore takes the value 0. In the third example we have even parity. Note the smooth transition from the start bit to the first data bit and from the parity bit to the beginning of the stop interval.

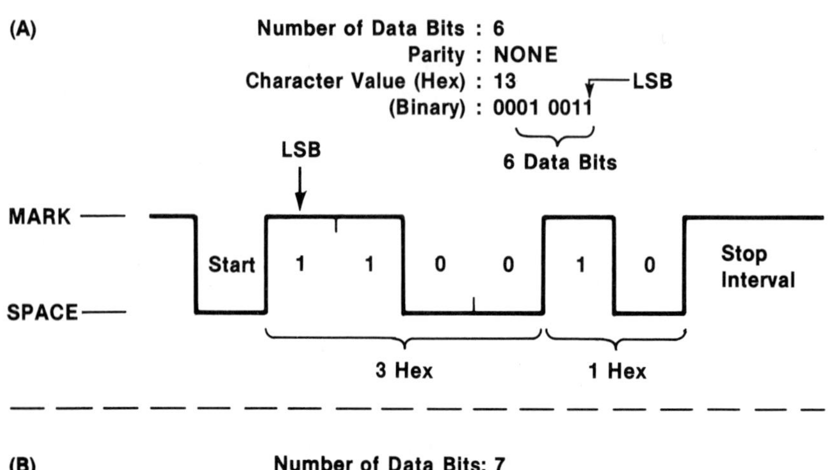

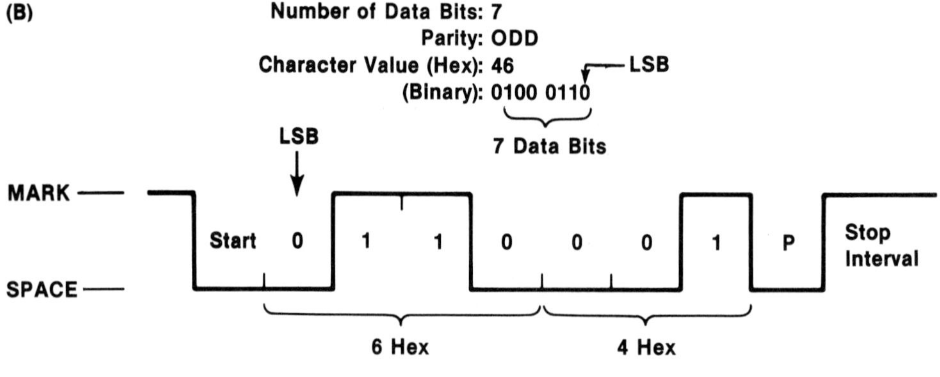

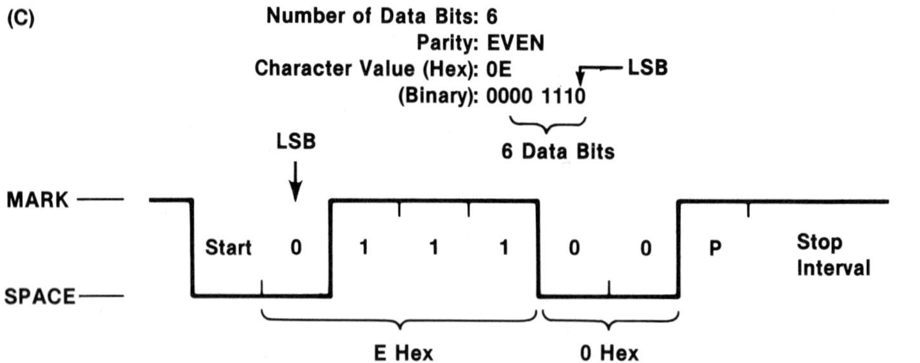

FIGURE 2–2 Asynchronous Character Waveforms

2·4 CALCULATION OF CHARACTER RATE

We are now in a position to calculate one of the most important numbers in data communications—the maximum rate, in characters per second, at which an asynchronous terminal can transmit. This maximum obviously happens when the transmitted characters are packed as closely as possible—that is, when the separation between each successive pair of characters is the minimum stop interval. The formula, in terms of bit rate and character size, is

$$\text{CHAR\_PER\_SEC} = \frac{\text{B\_P\_S}}{\text{BITS\_PER\_CHAR}}$$

We may compute the character size from

BITS_PER_CHAR = 1   (START BIT)

+ NUM_DATA_BITS

+ PARITY   (1 IF PRESENT,
                 0 IF ABSENT)

+ MIN_STOP_INTERVAL

The following is a typical specification for an asynchronous terminal or host computer port:

LINE SPEED:	300 bits/s
NUM_DATA_BITS:	7
PARITY:	Odd
MIN_STOP:	1 Bit Time

BITS_PER_CHAR:
$$
\begin{array}{rl}
1 & \text{(Start)} \\
+\ 7 & \text{(Data)} \\
+\ 1 & \text{(Parity)} \\
+\ 1 & \text{(Stop)} \\
\hline
10 &
\end{array}
$$

Do not forget to include the start bit in these calculations. It is usually not included in specifications because it is not a variable, but it is, nevertheless, present. The character rate is therefore

$$\text{CHAR\_PER\_SEC} = \frac{300}{10} = 30 \text{ char/s}$$

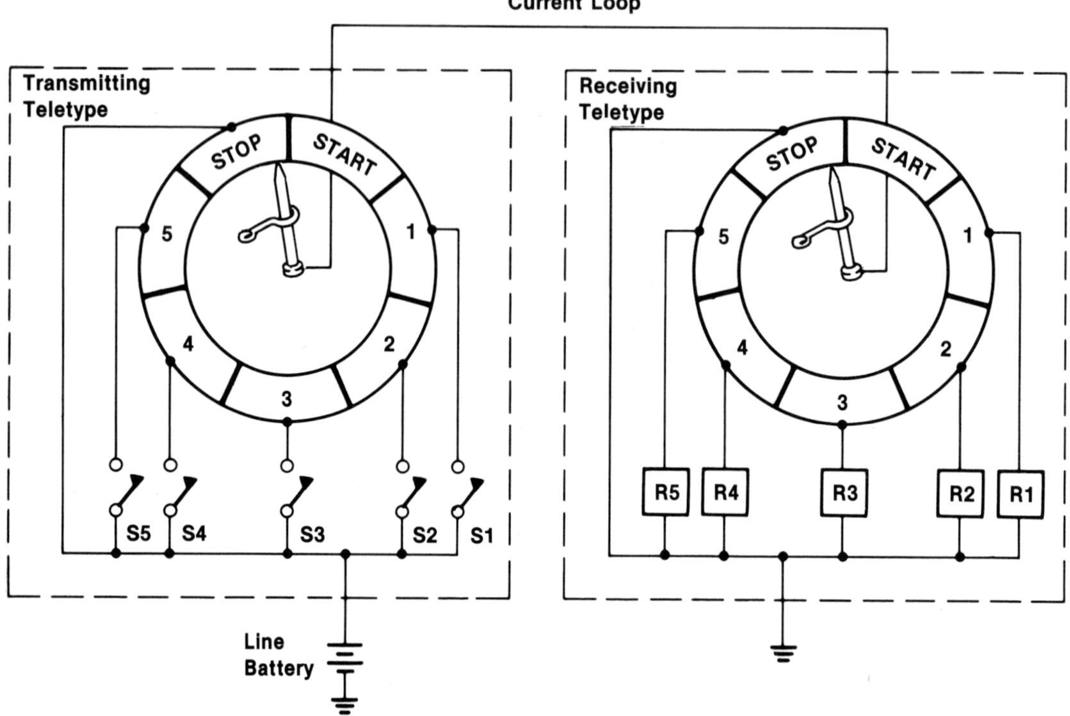

FIGURE 2–3   Baudot Distributor

2·5   AN ANTIQUE

Before we discuss the electronic devices that generate and receive asynchronous characters, let us look at the operation of an electromechanical device that performed the same function for many years before anyone dreamt of solid state circuits. This device is the Baudot (baw-DOUGH) distributor, named after the Frenchman who invented it about 1880 for use on teletype equipment. The device is illustrated schematically in Figure 2–3.

Each device (transmitter and receiver) is a ring of electrically conducting segments, insulated from each other. These segments represent the components of an asynchronous character with 5 data bits. There is an electrically conducting wiper in contact with each ring. The wiper in the transmitting teletype is connected to the wiper in the receiving teletype by a communications circuit capable of carrying direct current.

In both teletypes, the wipers are driven by motors which attempt to rotate them in a clockwise direction. The rotation, however, is prevented by mechanical devices symbolically represented by the "hooks." Note that the wipers are slightly to the left of top center and in contact with the *stop* elements. The stop element in the transmitter is directly connected to the line battery and grounded in the receiver, thus allowing current to flow in the line. In this system, called a *current loop,* current flow is interpreted as MARK, and absence of current is interpreted as SPACE.

The source of data in the transmitter is either a keyboard or a paper tape with punched holes representing the codes of the data characters. In either case, when a character is ready to be transmitted the switches (S1–S5) remain open or closed in correspondence with the data bits of the character. At this time the hook releases, allowing the wiper to rotate. As soon as the wiper crosses from the stop element to the *start* element, current flow stops (the line will go to SPACE). Not shown in Figure 2–3 is a mechanism in the receiver teletype which detects this current transition and releases the hook holding the receiver wiper. Since both motors are adjusted for the same speed, the wipers rotate in synchronism and enter *segment one* at the same time. At this point, transmitter switch S1 (which represents the least significant bit of the transmitted character) is directly connected to a relay coil (R1) in the receiver. R1 therefore latches into one state or another on the basis of the condition of S1. This continues as the wipers rotate, and by the time they leave *segment five,* R1–R5 contain copies of the states of S1–S5. A mechanism in the receiving teletype positions a printing device on the basis of the settings of R1–R5, and data transmission is complete. As the wipers approach the end of the stop elements, they are recaptured by the hooks, and we are ready for the next transmission cycle.

This is a fascinating device and a great accomplishment for its time, but it leaves room for improvement. It is subject to mechanical wear, and proper operation strongly depends on the motors maintaining equal speeds. It would be very difficult to adjust this device for a different number of data bits or to add a parity bit. It is also slow by electronic standards. Early teletypes operated at 50 bits/s; more modern ones operate at 110 bits/s. In the next two chapters we will see how electronic technology has dealt with these problems.

An interesting spin-off of this subject is the origin of the term *break.* You may have noticed a key marked break on some terminals and personal computers. A break occurs when a communications line goes to the spacing condition and remains there for at least one full character time. That is,

$$BREAK\_DURATION = BITS\_PER\_CHAR * T$$

In a system where SPACE is defined as the absence of current flow, the origin of the term is obvious. A very long SPACE occurs when the current loop is physically broken.

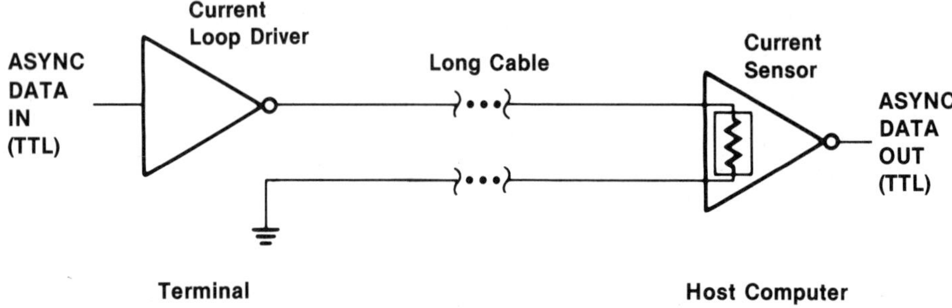

FIGURE 2–4   A Current Loop Interface

2·6   DISTORTION

Current loop circuits are still commonly used to connect terminals to host computers in the same building. It is an inexpensive alternative to modems in situations where the distance is just a little too long for digital circuits. The line drivers and receivers are now electronic rather than mechanical. A circuit of this type is shown in Figure 2–4.

The first waveform in Figure 2–5 shows typical data as it appears at the input to the current loop driver. Its characteristics are TTL voltage levels, square edges, and all bits of equal duration. The second waveform is an example of what we might observe at the input to the current sensor. Several effects are present:

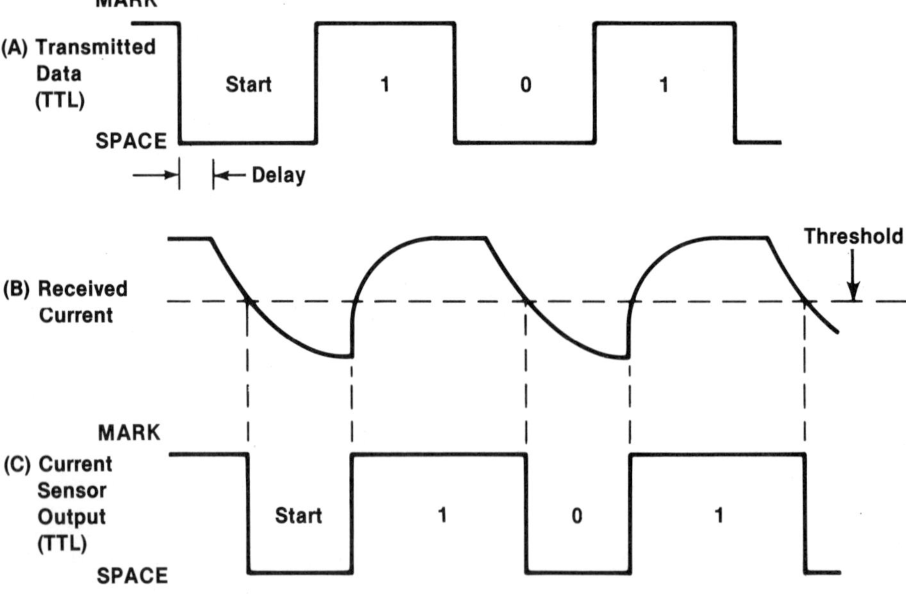

FIGURE 2–5   Character Distortion

- There is a delay which is partly due to transmission down the cable and partly due to the driver circuit.
- The waveform has become rounded because of cable inductance and capacitance.
- Because of the design characteristics of the driver, the rounding is more pronounced on falling edges than on rising edges (the opposite may also be true).

Rounded waveforms are troublesome to digital circuits. The current sensor is therefore designed to restore the waveform to TTL levels and sharpen the edges by establishing a threshold. The current sensor's output (the third waveform in Figure 2–5) is determined on the basis of the instantaneous difference between the received current and the threshold. Note that the timing of the waveform has not been perfectly restored. The marking pulses have been lengthened at the expense of the spacing pulses. Any such displacement of waveform edges from their proper places relative to the start bit is called *teletype distortion*.

Teletype distortion has many causes, and it can occur on modem driven phone lines as well as current loops. Among the causes are

- noise
- nonlinearities in drivers or transmission medium (the difference between rise and fall times in Figure 2–5B is an example of a nonlinearity)
- speed differences between transmitter and receiver

The case of a speed mismatch is particularly bad because the effect adds from bit to bit over the length of the character.

There is a standardized way to measure teletype distortion. Assume that we have transmitted a 4-bit character (1001 binary) at 200 bits/s and the waveform in Figure 2–6 is observed on an oscilloscope at the receiver. If we make the

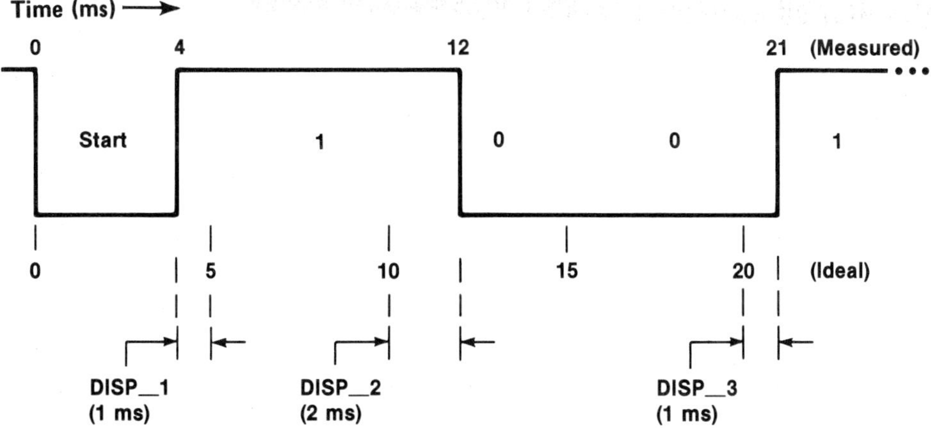

FIGURE 2–6 Measurement of Character Distortion

leading edge of the start bit our reference point, the measured times (in milliseconds) to each edge of the waveform are indicated at the top of the diagram. Distortion is obviously present—but how much?

Since we know that we are dealing with a 200-bit/s signal (T = 5 ms), we can compute where the bit transitions should ideally be (if there was no distortion). We have marked these below the waveform. We can now scan the waveform and measure the displacement of each edge from its ideal position. The end of the start bit, which should occur at 5 ms actually happens at 4 ms. The end of the first data bit is displaced by 2 ms, and so on. Between the second and third data bits we obviously cannot determine if any displacement is present because there is no change of polarity. Also note that our displacements are absolute values. For this calculation, it does not matter whether the displaced edge occurs too early or too late.

We now look at the collection of displacements in Figure 2–6 and select the largest one (DISP_MAX). This displacement, expressed as a percentage of the bit time, is the standard measure of character distortion.

$$\text{CHAR\_DISTORT} = \frac{\text{DISP\_MAX}}{\text{T}} * 100$$

$$= \frac{2 \text{ ms}}{5 \text{ ms}} * 100 = 40\%$$

REVIEW QUESTIONS

1. Sketch the waveforms of the asynchronous characters specified. Label all bits and indicate the time of occurrence of each bit transition (assuming the leading edge of the start bit is T = 0).

	NUMBER OF DATA BITS	PARITY	CHARACTER VALUE (HEX)	LINE SPEED (BITS/S)
a.	8	none	D5	1000
b.	6	even	26	500
c.	7	odd	7F	100
d.	6	even	22	200
e.	5	none	15	50

2. Compute the maximum number of characters per second that can be transmitted by asynchronous terminals with the following specifications:

	NUMBER OF DATA BITS	PARITY	STOP INTERVAL	LINE SPEED (BITS/S)
a.	7	odd	2	110
b.	5	none	1.5	50
c.	8	none	1	9600

3. An asynchronous terminal with a nominal line speed of 200 bits/s is transmitting too fast by a factor of 5%. At the end of the seventh data bit, how large is the time displacement from the ideal bit transition point to the bit transition point that this malfunctioning terminal will produce? What is the distortion in percent?

4. Compute the minimum duration of a break on an asynchronous line specified as follows: 6 data bits, odd parity, 2-bit stop interval, 300 bits/s.

5. A character whose data bits are all 1s is received on a 1000-bit/s asynchronous line. A scope measurement shows that the duration of the start bit is 0.8 ms. What is the character distortion in percent?

6. The character shown is received on a 100-bit/s asynchronous line. Bit transition times are indicated above the waveform (in ms).

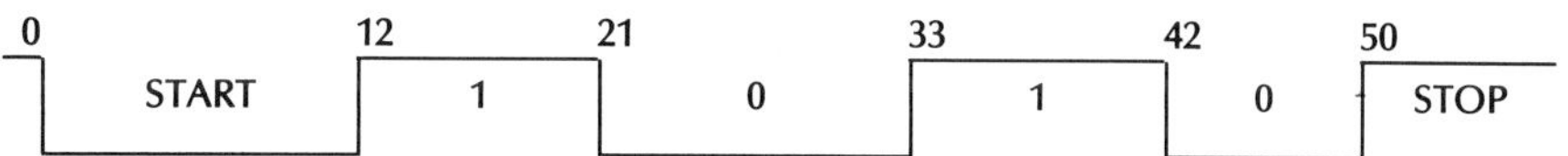

Compute the displacement of each transition point from its ideal location. Identify the largest displacement and compute the character distortion in percent.

chapter three

THE UART

The Universal Asynchronous Receiver/Transmitter (UART) is an electronic intermediary between a microprocessor and a communications line and therefore falls into the category of input/output (I/O) devices. To fully understand these devices, we will digress to review the functions of various hardware elements in a small computer. Because technical subjects are easily rendered confusing by excessive detail, the picture has been highly simplified, but the concepts are applicable to most systems. If you are already an expert on small computer hardware, feel free to skip or skim to Section 3•2.

A Typical Personal Computer System
Courtesy of International Business Machines Corporation

3·1 PROGRAMMED I/O IN A SMALL COMPUTER

Figure 3–1 illustrates the fundamental architectural components of a small computer. There are three major types of LSI (large scale integration) chips that go into these systems.

Microprocessor

There is only one microprocessor chip in a small computer, and it controls all movement and processing of data in the system. The available processing operations include all sorts of arithmetic and logical functions which may be applied to data items of various sizes. The processing takes place according to a list of instructions (a *program*) that are written by a programmer and then

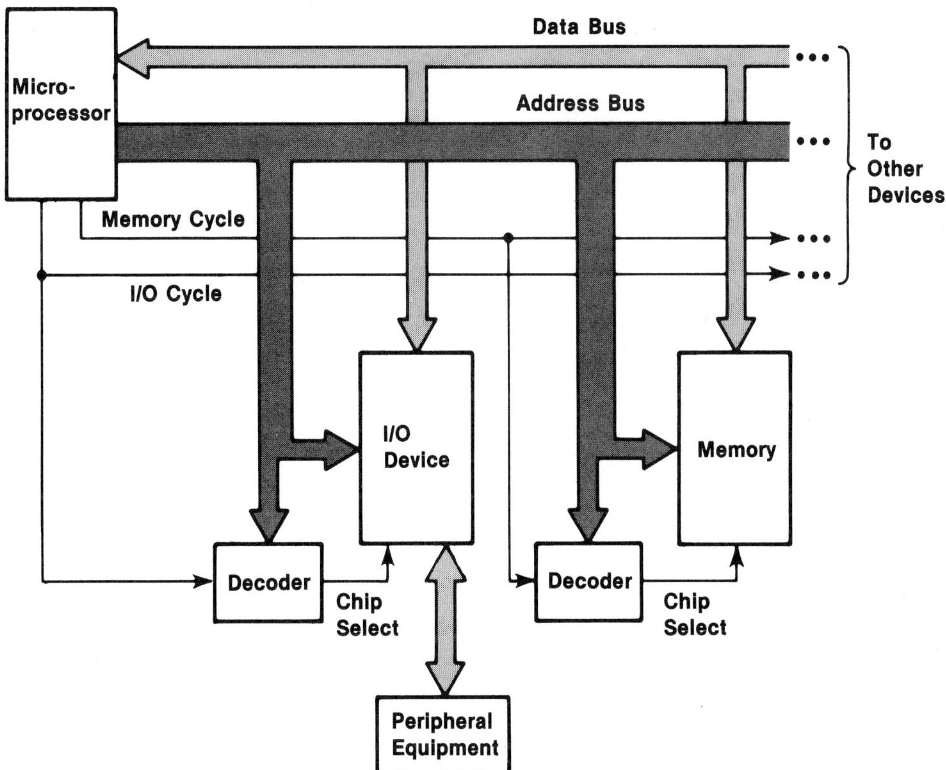

FIGURE 3–1 Small Computer Hardware Architecture

stored in memory. These instructions are usually executed in the sequence in which they are stored, except for the two circumstances described as follows:

BRANCHING A microprocessor may be instructed to compare the values of two different data items and, on the basis of the result (equality or inequality; less than or greater than), cause the program to jump to some instruction other than the one immediately following. This ability to *branch* allows a computer to go back and repeat a group of processing steps for a number of items in a list or to make "decisions" based on information provided by an operator. Most of the apparent "intelligence" of computers stems from their ability to make conditional branches.

SUBROUTINES Frequently, a group of instructions must be repeated at many different places in a long program. The programmer may then decide to isolate and assign a name to this set of steps. This named group of instructions is a *subroutine*. At any point where the subroutine's function is required, the programmer will insert a single instruction which will *call* the subroutine by name. The microprocessor will then execute the instructions in the named subroutine. At the end of the subroutine there will be a *return* instruction which will cause the program to jump back to the instruction immediately following the latest call (whose location was conveniently saved by the microprocessor at the time of the call).

Memory

A small computer usually contains several memory chips (only one is shown in Figure 3–1). A memory device is a large collection of *cells,* each of which can store a data item (typically 1 byte). Each cell is identified by a binary number called an *address.* Data transfer to and from memory is totally under the control of the microprocessor. Transfers from the microprocessor are called *write cycles;* transfers in the opposite direction are called *read cycles.* Most of the instructions in the microprocessor's set will result in data transfers to or from memory.

These data transfers take place via *busses*—groups of parallel wires carrying related signals. When the microprocessor wants to write to memory, the data byte is placed on the *data bus,* and the *address bus* specifies the cell that will store the data. Since the computer may contain several memory devices, the address is split: the least significant part specifies the cell within a device, and the most significant part specifies the desired device. Each device has a *decoder* which is set to recognize the unique binary pattern for its associated device. When this pattern is present on the address bus the decoder asserts a wire called *chip select* which activates its connected memory device. If its chip select is not asserted, the memory device remains dormant.

I/O Devices

From the microprocessor's point of view, an *I/O device* looks much like a memory. Rather than cells, the data storage elements in I/O devices are called *registers*. The number of registers in an I/O device is usually far smaller than the number of cells in a typical memory chip, but the addressing and chip selection mechanisms are identical. In most microprocessors the number of instructions that transfer data to and from the I/O device is small. In some cases there are only two basic I/O instructions: IN for read cycles; OUT for write cycles. The one additional complication shown in Figure 3–1 is the *I/O cycle* and *memory cycle* wires. In most computer systems there are memory cells and I/O registers sharing the same address. The microprocessor, on the basis of the type of instruction it is executing, knows exactly which device it wants and will assert either the I/O cycle line or the memory cycle line to announce its desires. These wires must therefore enter the decoders and participate in the chip selection process.

From the user's point of view, I/O devices are strikingly different from memories because each I/O device is attached to a piece of *peripheral equipment*. Peripherals are "outside world" devices including communications lines, printers, keyboards, and disks. The I/O registers allow the microprocessor to interact with the outside world. For example, data written into I/O registers can cause a character to be sent on a communications line or position the head of a disk drive to a new track. Outside events (the arrival of a new character from a communications line, for example) can change the contents of an I/O register, and news of the external event is therefore available to the microprocessor. Having the news available does not mean the microprocessor is forced to read it. If the program does not include instructions to frequently read I/O registers, events will pass unnoticed. This type of interaction is therefore called *programmed I/O*. Later in this chapter we will examine a technique which can force the microprocessor to take notice.

I hope that the reasons for this long digression are now apparent. In any case, we are now ready to study the internal operation of a UART. We will start with some generalizations and proceed, in the next chapter, to a specific example.

3·2   UART RECEIVER FUNCTIONS

The UART is composed of two independent sections: a receiver and a transmitter. It is therefore a full-duplex device. We will look at the receiver first. The major functions of the UART receiver are

- start bit detection
- serial-to-parallel conversion
- error detection

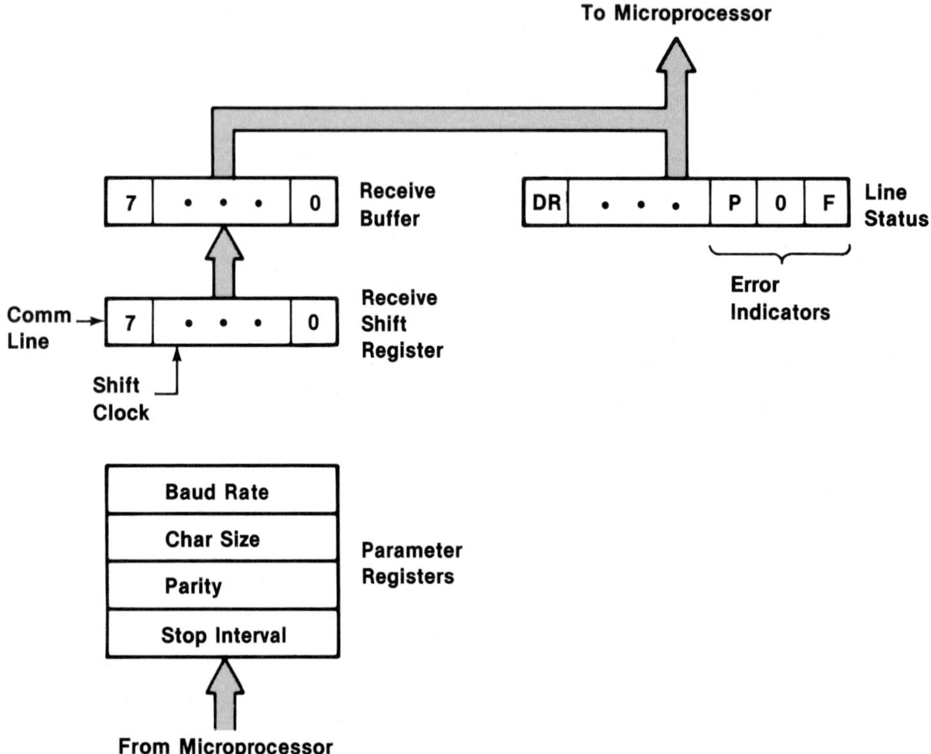

FIGURE 3–2   UART Receiver Registers

The registers involved in the receive function are shown in Figure 3–2. There is a great deal of additional control circuitry that is not shown. The *Receive Buffer* and *Line Status Registers* are readable by the microprocessor. The *Parameter Registers* can be loaded by the microprocessor with values that determine what format the receiver will expect for characters on the incoming line. The values that we set for baud rate, parity, character size, and stop interval must match the actual format on the line. The ability to adjust these parameters from the microprocessor's program allows us to use the same hardware for connection to many different devices. In the next chapter we will see some specific examples of the possible settings for these format parameters.

The input from the communications line appears at the left side of the *Receive Shift Register* (which is neither readable nor writable by the microprocessor). This is where serial-to-parallel conversion takes place. Shift clock pulses (see Figure 3–3) must be generated as closely as possible to the center of each bit. This requires a delay of T/2 from the beginning of the start bit for the first pulse, followed by delays equal to T for the remaining data bits (and the parity

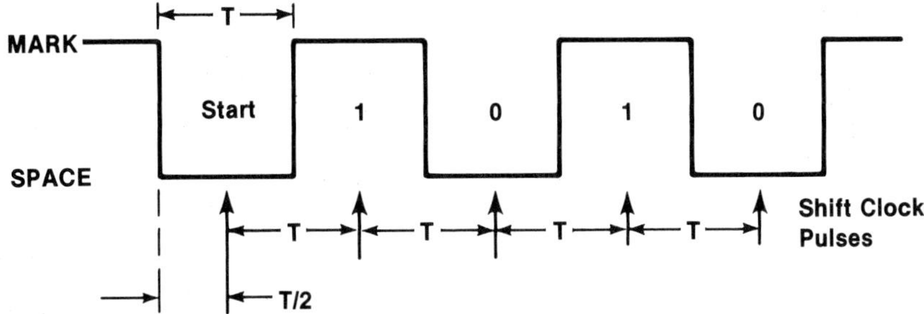

FIGURE 3–3   Receiver Shift Timing

bit if we are expecting one). In the next section we will see exactly how these shift clocks are generated. The number of clocks generated will depend on the contents of the character size and parity Parameter Registers.

After the last data bit has been shifted in, control circuits transfer the character, in parallel, from the Shift Register to the Receive Buffer and, simultaneously, the DR (Data Ready) bit in the Line Status Register will be set to 1. DR is a "flag" whose purpose is to notify the program in the microprocessor of the arrival of the new character. The program, if it has been paying attention (periodically reading the Line Status Register), will fetch the character from the Receive Buffer and usually store it in memory for later processing. When the Receive Buffer is read, the DR flag will be cleared (reset to 0) to indicate that the Receive Buffer is once again empty. This prevents the program from fetching the same character twice.

The technique described in the preceding paragraph is called *double buffering,* and its purpose is to provide the program with extra time to respond to DR. We could eliminate the Receive Buffer and allow the program to directly read the character from the Shift Register. At the end of the minimum stop interval, however, a new character may appear on the line and disturb the contents of the Shift Register. The time between the program's periodic checks on the condition of DR must therefore be less than the minimum stop interval. By keeping the new character in the Receive Buffer, we can guarantee that it will remain undisturbed for at least the full duration of a character on the line. This substantially reduces the frequency at which the program has to check DR.

In Figure 3–4 we have a flowchart presenting a more detailed view of the UART receive operation, including the error detection function. We begin at the top, assuming that the line is in the idle interval between characters. The line is therefore at MARK, and nothing happens until we see a transition to SPACE. We suspect this event to be the beginning of a start bit, but we must be certain. The MARK-to-SPACE transition could have been due to a short noise spike or "glitch," and we do not want to trigger the shift clock generation circuits on a glitch. We therefore wait one-half bit time (T/2) and test the state of the line again. T/2 is measured in milliseconds (or at least hundreds of microseconds) and noise spikes are usually a few microseconds or less. Therefore, if

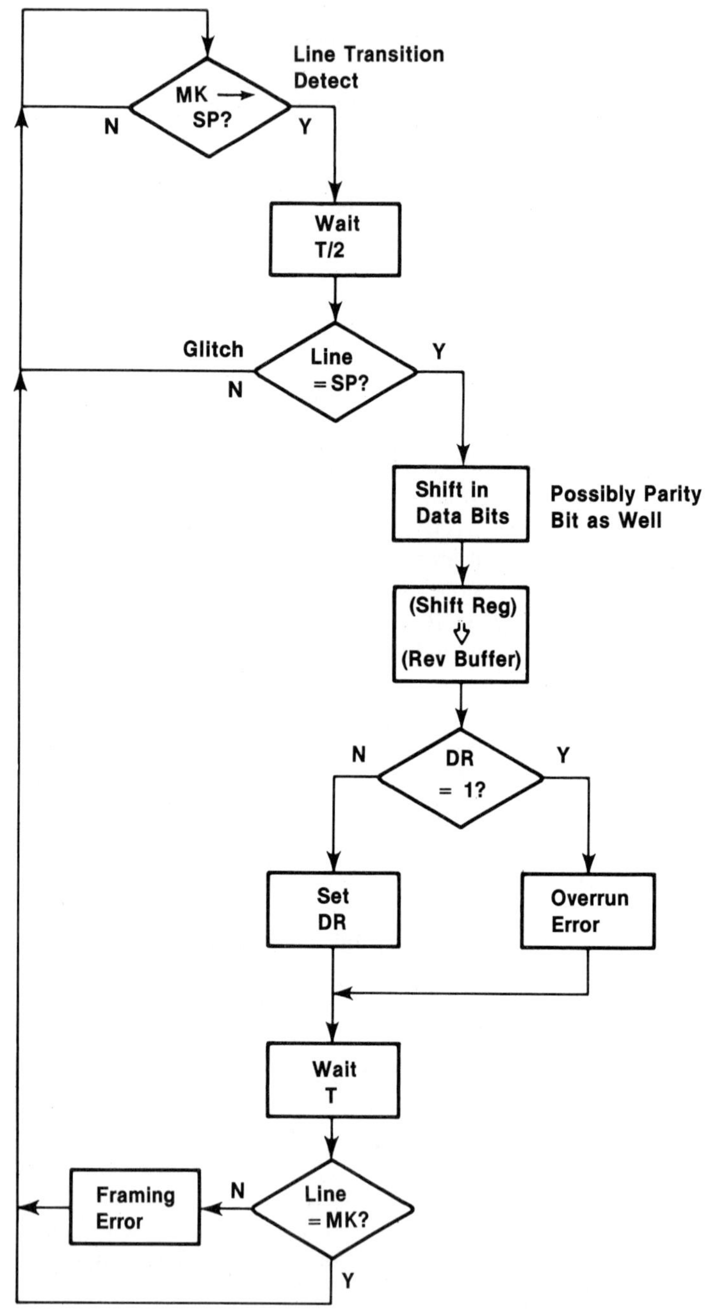

FIGURE 3–4   UART Receiver Functional Flowchart

the line is still in a spacing state after T/2, we can be reasonably certain that we are dealing with a real start bit. Otherwise, we declare this to be a false start and return to the stop of the flowchart.

If we are convinced that the start bit is real, the shift clock generator will be triggered, and the appropriate number of data bits will enter the Shift Register. If we are configured to expect a parity bit, it will also be shifted in, and the UART circuits will test the number of 1s in the received character. If the count does not comply with the parity rule specified in the Parameter Register, the P error indicator will be set in the Line Status Register.

The shifting process is now complete, and the new character is transferred to the Receive Buffer. At this point the UART control circuits test the state of the DR flag. If it is already set the program has been asleep. The previous character in the Receive Buffer has not been read into the microprocessor, and we have just destroyed it with the new character from the Shift Register. This is an *overrun error,* and it causes the O error indicator to be set. If we do not have an overrun condition, DR will now be set.

We then wait one more bit time (which carries us into the stop interval) and test the state of the line once again. If we are indeed into a stop interval, the line state should be MARK. If it is not, we declare a *framing error* and set the F error indicator. Framing errors can occur if we turn on the UART in the middle of the steady stream of incoming characters. In this case we may not enter the top of the flowchart during an idle interval, and, thus, we may mistakenly assume that a spacing data bit is a start bit. This will begin the shifting process in the middle of a character and possibly end it in the middle of the following character. We can also get framing errors if the baud rate setting for the receiver is not correct. Can you explain why?

In any case, we have done all the possible processing for a received character and are ready to return to the top of the flowchart and start again.

3·3 RECEIVER SHIFT TIMING

In Figure 3–3 we have illustrated the timing relationships between the leading edge of the start bit and the receive shift clocks. Figure 3–5 is a simplified logic diagram of the circuit generating these clocks. Notice that the clocks are separated by the bit time T and are therefore repeating at a frequency equal to B_P_S. This separation is simple enough to generate. Suppose we have a square wave, called the "fast clock," whose frequency is exactly 8 times the bit rate of the line.[1] We apply this clock to the input of a 3-bit counter that will divide the frequency by 8. If we decode one of the counter's eight states (we assume the chosen state is 000), the decoder output pulses will be separated by exactly T

---

1. As we will see, the ratio of the fast clock to the bit rate of the line is an option which may be selected by the designer of the UART. For this first example we will use a value of 8.

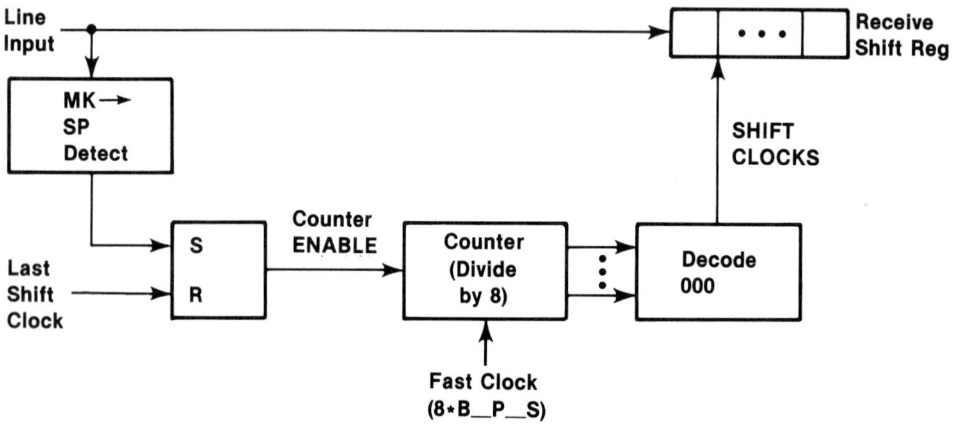

FIGURE 3–5   Receive Clock Generation Logic

and are suitable for use as shift clocks. Now that we have the correct separation, we must obtain the T/2 delay between the leading edge of the start bit and the first shift pulse.

The counter in Figure 3–5 has an ENABLE input in addition to the fast clock. When ENABLE is false, the counter is forced into state 4 (100 binary) and held in that condition. When enable becomes true the counter is allowed to change state in response to the fast clock input. ENABLE is controlled by the output of a flip-flop. For the moment, assume that the flip-flop output is 0 and the communications line is in an idle interval. The upper waveform in Figure 3–6 shows this situation at the left side of the drawing. The line is in the marking condition, and the ticks of the fast clock do not change the counter state.

When the line goes from MARK to SPACE, the flip-flop becomes asserted. In the upper waveform this happens immediately before a tick of the fast clock. The counter almost instantly goes to state 5, then to states 6, 7, and rolls over to state 0. State 0 is the one that will produce our shift clock. Assume that the shift register is designed such that the actual shifting takes place on the trailing edge of the clock pulse. As we can see in Figure 3–6, the shift will take place precisely at the midpoint of the start bit. Since we have already established that the clocks are separated by exactly one bit time, all other bits will be clocked precisely at mid-bit too.

We have obtained our desired T/2 delay from the leading edge of the start bit. This works for the following reason: when the counter was not enabled we forced it to a state (4) that was at the halfway point in the counting sequence between the states we chose for the shift decode. This forced counter state is called a *preset*. There are other possible ways to design this logic. Suppose we started with a fast clock at 10 times B_P_S rather than 8. We would have to redesign the counter to divide by 10, and the halfway point in the count sequence would now be 5 instead of 4.

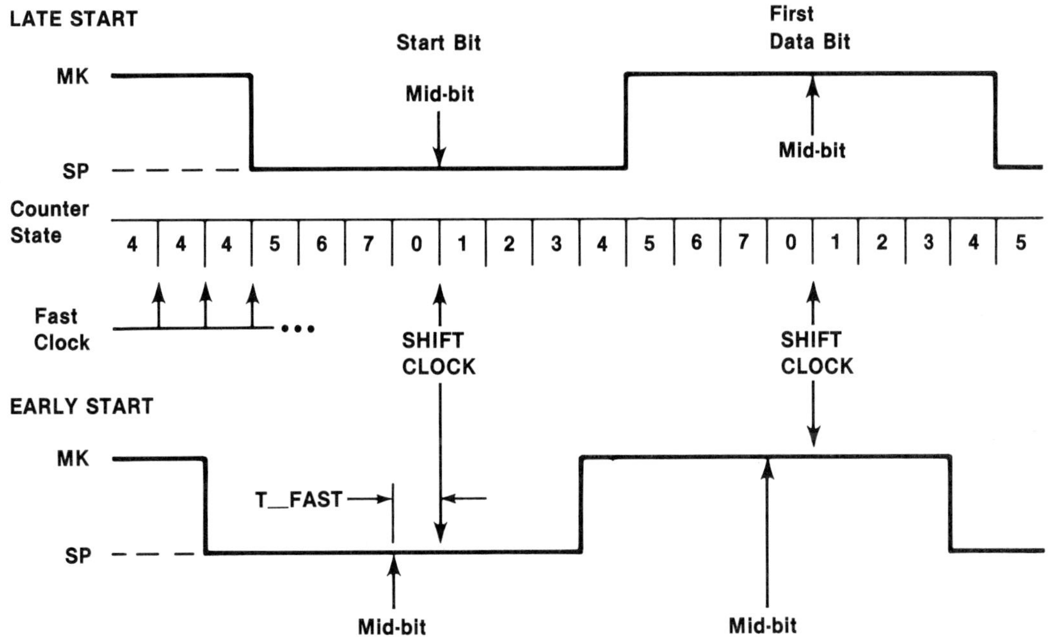

FIGURE 3–6 Shift Clock Timing Variations

It appears that we have an excellent clock generation scheme. However, let us look at what happens when the leading edge of the start bit occurs just after a tick of the fast clock (the lower waveform in Figure 3–6) rather than just before the tick, as it did in the upper waveform. The start bit has "missed the boat," and we must wait almost a full period of the fast clock before the counter experiences the first state change. As a result, the shift occurs after the mid-bit point. The time by which the shift is late is exactly equal to one period of the fast clock (T_FAST). By definition of an asynchronous line, we have no control over the relationship between start bits and fast clock ticks, and, for any given character, we have a situation that lies somewhere between the extremes illustrated in Figure 3–6. We can reduce this uncertainty by increasing the frequency of the fast clock and thereby dividing the bit time into smaller intervals. Of course, this also requires that we increase the counter's frequency division ratio to compensate.

What is the importance of this sliding about of the shift point? It reduces the amount of character distortion the UART can tolerate without making errors. In the lower waveform of Figure 3–6 the shift point moved closer to the trailing edges of the character's bits. If the transition at the end of the first data bit was sufficiently distorted, we might erroneously shift in a SPACE instead of MARK. If the shift point is exactly at mid-bit, we can have distortion up to 50% before errors occur. This is the best we can ever do. As the shift point moves closer to the bit transition point, this tolerance is reduced. The amount by which it is reduced is the percentage of a bit time (T) that is represented by the period of the fast clock (T_FAST). In algebraic notation:

$$\text{DISTORT\_TOLERANCE} = 50 - \left(100 * \frac{\text{T\_FAST}}{\text{T}}\right)$$

We can also express this in terms of B_P_S and the frequency of the fast clock (F_FAST):

$$\text{DISTORT\_TOLERANCE} = 50 - \left(100 * \frac{\frac{1}{\text{F\_FAST}}}{\frac{1}{\text{B\_P\_S}}}\right)$$

$$= 50 - \left(100 * \frac{\text{B\_P\_S}}{\text{F\_FAST}}\right)$$

This is a more convenient form since the frequency of the fast clock is usually specified in terms of B_P_S. For example, what is the distortion tolerance of a UART receiver operating with a fast clock of 16 times a bit rate?

$$\text{DISTORT\_TOLERANCE} = 50 - \left(100 * \frac{\text{B\_P\_S}}{16 * \text{B\_P\_S}}\right)$$

$$= 50 - \left(100 * \frac{1}{16}\right) = 43.75\%$$

As a final comment, notice that the uncertainty in the shift point fell entirely after the mid-bit point. This happened because we assumed the shift point to be the trailing edge of the decoded counter state. How could we modify the design to make the uncertainty fall before the mid-bit point? Could we make the range of uncertainty straddle the mid-bit point?

3·4   UART TRANSMITTER FUNCTIONS

We now turn to the other half of the UART. A simplified logic diagram of the transmit section is shown in Figure 3–7. Two registers are involved: the *Transmit Hold Register*, which is loadable by the microprocessor, and the *Transmit Shift Register*, which is not accessible to the microprocessor. Once again we have double buffering. There is a series of gates between the two registers which provide a path for the parallel transfer of a character from the Hold

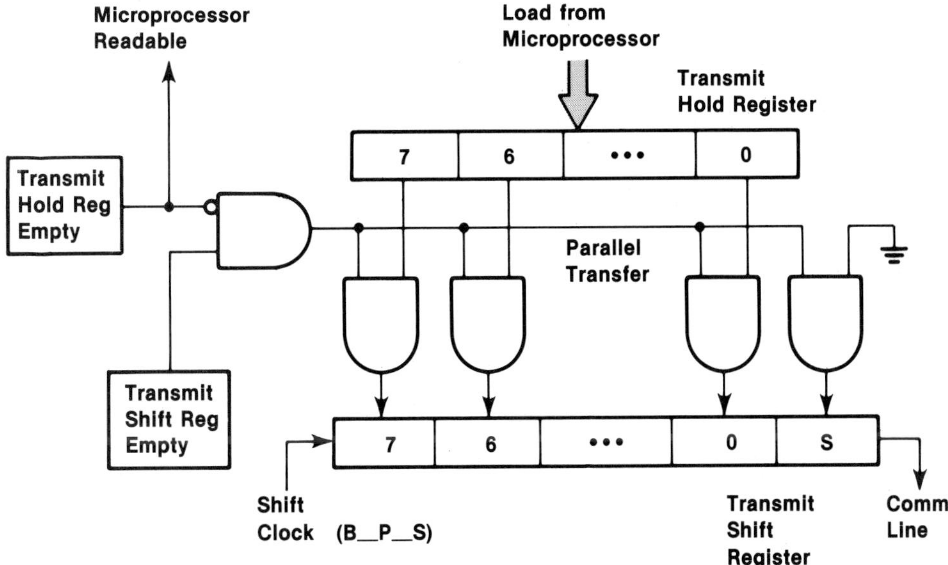

FIGURE 3–7   UART Transmitter Block Diagram

Register to the Shift Register. Note that whenever a character is transferred into the Shift Register, we also load a 0 into the start bit position at the right end of the Shift Register. Associated with each register is a 1-bit flag which is asserted when its respective register becomes "empty." The parallel transfer will take place when the output of the gate in Figure 3–7 becomes true. This happens when we have Hold Register NOT Empty and Shift Register Empty.

To see how this mechanism works let us follow the sequence of events that take place when three sequential characters are transmitted. The timing is illustrated in Figure 3–8. The transmit hold-register-empty flag is readable by the program. From the program's point of view the operation is simple: if the flag indicates that the Hold Register is empty, then it may be loaded with a character for transmission.

Assume that both registers are empty and the line is in an idle state (we are transmitting a MARK). The program has a 3-character message (char 1, char 2, char 3) that it wants to transmit. Since the hold-empty flag is true, char 1 can be loaded immediately. The Hold Register is no longer empty, but the Shift Register still is, and we have therefore satisfied the conditions for activating the parallel transfer. This causes the hold-empty flag to pop up once again. The program can now load char 2 into the Hold Register. Notice that the time between the first two loads by the program was rather short. Once again, the Hold Register is not empty (it contains char 2), but we will not immediately generate a second parallel transfer because the Shift Register empty condition is false.

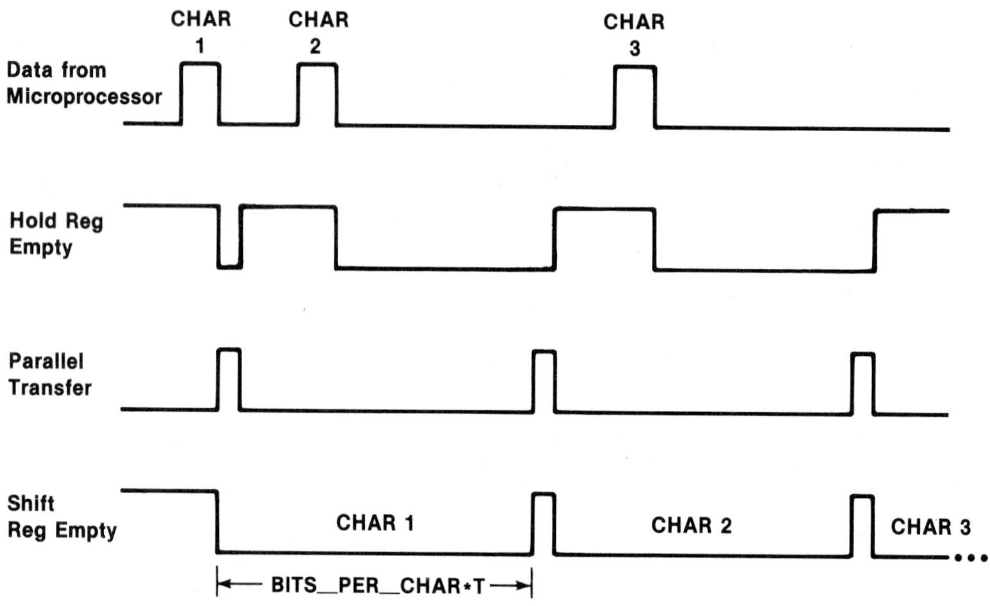

FIGURE 3–8   Transmitter Timing

When the first parallel transfer loaded char 1 into the Shift Register, we also triggered a series of shift pulses which will accomplish the parallel-to-serial conversion of the character. The separation between the clocks is bit time T, and the shifting process obviously lasts for a full character time (BITS_PER_ CHAR * T). Char 2 will remain "on hold" for all this time, and, as a result, the program must spend all of its time repeatedly testing the hold-register-empty flag while it is waiting for the opportunity to get rid of char 3. For the program this is an eternity. Microprocessor instruction times are usually a few microseconds. Character transmission times, even at the highest baud rates, are at least 1 ms.

When the shifting process for char 1 is finally complete (at the end of the minimum stop interval), the Shift Register becomes empty once again, and we get a second parallel transfer (char 2 into the Shift Register). The Hold Register is now available for char 3. The program senses this, and char 3 is soon on its way. Longer messages are sent by repeating these steps.

If all this is slightly confusing, think of it as a baseball game. The character in the Shift Register is at bat; the one in the Hold Register is on deck; the remainder of the message is in the dugout (memory). The program manages the team. (There is one peculiar rule: the first man at bat is required to momentarily pause in the on-deck circle on his way to the plate.)

3·5 IMPROVING SOFTWARE PERFORMANCE

By following the steps illustrated in the previous section, we observe that the process of transmitting a multicharacter message takes all of the microprocessor's attention, but most of the time is spent in nonproductive work (repeatedly testing the condition of the hold-register-empty flag). Let us hypothesize a mechanism that can eliminate all this flag testing. Suppose that every time the hold-register-empty flag became true, the UART could force the microprocessor to call a particular subroutine whose location was somehow specified by the UART. We could then proceed as follows:

- At a series of locations in memory, the program sets up the characters of the message to be transmitted and, at some other prearranged location, stores a count of the number of characters in the message.
- The program sends the first character to the UART transmitter. This will clear (negate) the hold-register-empty flag. The program then decrements the character count by one (because a character has been sent). The program can then go off to attend to some totally unrelated business (we will see examples shortly) and not concern itself with this message anymore.
- When the hold-register-empty flag becomes true, the subroutine call will be forced by the UART. Instructions in the subroutine will fetch the next character of the message from memory; load it into the UART transmitter; decrement the character count; return the microprocessor to whatever instruction it was executing just before this interruption took place.
- The previous step is repeated until the entire message has been transmitted.

A mechanism such as the one just described does, of course, exist. It is called an *interrupt*. The definition is important enough to repeat for emphasis:

An *interrupt* is a subroutine call that is forced to occur by an event external to the microprocessor.

There are numerous other examples of events that cause interrupts. The receive side of the UART may cause an interrupt when the Data Ready flag becomes true. Other examples are the arrival of a disk drive head at a new track; the completion of a line of text by a printer; a door or window opening in an intrusion alarm system. The actual mechanism is usually a pin on the microprocessor chip which must be asserted by the interrupting device. Associated with the interrupt is a number called a *vector*, which is passed on the data bus by the interrupting device. The vector specifies the address of the subroutine that will run. The subroutine itself is usually called an *interrupt service routine*.

Now that the program has all this free time on its hands, how can we use it? In a multiuser system, we can turn the control of the microprocessor over to another user's program. This is the basis for time-sharing. In a single-user system, such as a personal computer, there is frequently nothing else to do, and interrupts therefore do not add much to system performance. There are a few examples of personal computer programs that take advantage of this feature. Some word processing software allows the operator to edit one file while a second file is being printed. Personal computer printers are frequently connected to asynchronous serial ports driven by UARTs. In this situation, the word processor program would get the file for printing from the disk, set it up in memory as described earlier, and initiate transmission of the first character. The remaining characters will be transmitted in response to the interrupts while the word processor program continues to accept editing commands from the keyboard. The operator (assuming no distraction by the sound of the printer) does not notice any difference except for a slight slowdown in the word processor's response to the commands.

REVIEW QUESTIONS

1. Could we design a UART transmitter without a Transmit Hold Register? What impact would this have on the program transmitting the data?

2. Would an incoming line break produce a framing error in a UART receiver? Are all framing errors due to an incoming break? Explain your answer.

3. Draw the "early start" and "late start" waveforms (see Figure 3–6) for the following conditions:

	F_FAST	COUNTER DECODE STATE	COUNTER PRESET
a.	8 * B_P_S	1	4
b.	4 * B_P_S	0	2
c.	5 * B_P_S	0	3

   Assume that the shift clock takes place at the end of the decoded counter state. Show the time relation of the shift clock to the mid-bit point for all cases.

4. Measurements on an asynchronous line show that characters are being received with distortions up to 40%. If the UART receiver is operating with F_FAST equal to 8 * B_P_S, will this data be received without errors? What if F_FAST is increased to 16 * B_P_S?

5. A UART receiver that is configured for 6 data bits and even parity is connected to a transmitter that is configured for 8 data bits and no parity. Assuming that both are operating at the same baud rate, what errors might you expect to see in the receiver? Draw a sample waveform from this transmitter that would create each error type. If the transmitter configuration were changed to 7 data bits and no parity, would you still get the same error types at the receiver?

6. Describe the functions that would have to be present in the UART and its controlling program if we wanted to implement an interrupt mechanism for received characters.

chapter four

A REAL UART ON A REAL PCB

We will now examine an actual LSI chip design based on the principles discussed in the last chapter. The 8250 UART is produced by several well-known semiconductor manufacturers, and it is the foundation upon which asynchronous communications features are built into the IBM Personal Computer. In the IBM PC the 8250 appears on a small printed circuit board called the Asynchronous Communications Adapter (ACA), which we describe in Section 4•8.

The 8250 is contained in a 40-pin package. Figure 4–1 is a block diagram showing the pertinent interface signals. For the sake of simplicity we have not shown pins such as address and data strobes. There are also some pins whose functions are related to subjects that will be covered in later chapters, and these too are not shown.

There are three input pins (A0–A2) which are used for selection of the 8250's internal registers. These will form the least significant part of an I/O device address as described at the beginning of Chapter 3. The most significant part of the 8250's device address range will activate the chip select function. The 8250 provides three chip select pins. All of them must be simultaneously asserted to allow data transfer between the 8250 and the microprocessor. CS0 and CS1 are asserted by a high logic level (1), and CS2 is asserted by a low logic level. An 8-bit bidirectional (read/write) data bus is provided by D0–D7. This means that all registers in the 8250 will have to be 1 byte wide.

There are a few other miscellaneous input pins, including *master reset* and *crystal oscillator*. Activation of master reset puts all registers into a state in which most of the 8250's functions are disabled. Master reset is usually activated momentarily at the time AC power is turned on or, in personal computers, any time the system is "booted." The crystal oscillator provides a master clock signal for all the 8250's control logic circuits. Any precisely timed functions, such as shift clock generation or start bit detection, are derived from the crystal oscillator. In the IBM PC the oscillator operates at 1.8432 MHz. We will shortly see why this number was chosen.

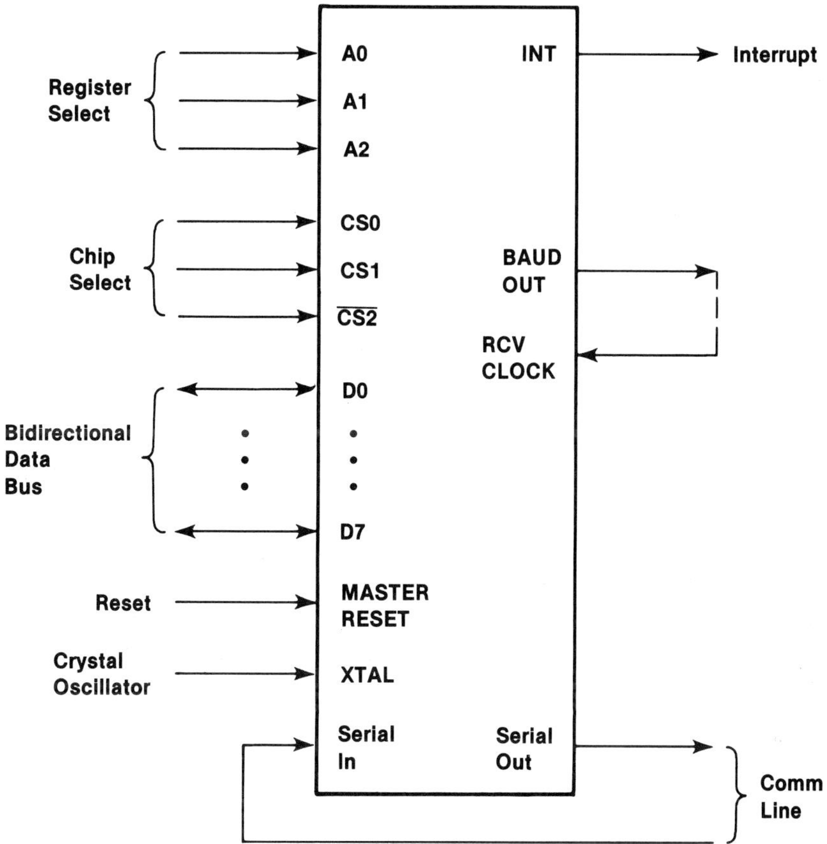

FIGURE 4–1   The 8250 UART

On the output side, the 8250 has a pin for initiating interrupts (INT). The 8250 does not generate the vector. In the IBM PC the vector is provided by another device (an interrupt controller chip) which is triggered by the INT line. There is also a pair of pins related to baud rate selection, which we will discuss later, and the communication line interface.

4·1   REGISTER ADDRESSING

The registers of the 8250 are listed in Table 4–1 along with their addresses. The functions of each register will be explained later.

For the moment ignore the DLAB and Cycle columns. The first thing we notice is that there are ten registers, and since three bits are provided for register selection, we have only eight possible addresses. In fact, three registers share

TABLE 4–1   Register Addressing

| REGISTER | ADDRESS (A2, A1, A0) | | DLAB | CYCLE |
	Decimal	Binary		
Receive Buffer	0	000	0	Read
Transmit Hold	0	000	0	Write
Div Latch (LS Byte)	0	000	1	
Div Latch (MS Byte)	1	001	1	
Interrupt Enable	1	001	0	
Interrupt Ident	2	010	-	
Line Control	3	011	-	
Modem Control	4	100	-	
Line Status	5	101	-	
Modem Status	6	110	-	

address 0, and two registers share address 1. The remaining five have unique addresses. This is a problem, for the program must be absolutely sure of which register it is going to access on a particular IN or OUT instruction.

The way out of this tangle is provided by a bit in the Line Control Register called DLAB (Divisor Latch Access Bit). If DLAB is set to 1, the shared addresses refer to the two sections of the Divisor Latch; if DLAB is 0, the shared addresses refer to other registers. DLAB fortunately is in a register (Line Control) which has a unique address, and the program can therefore adjust its value at any time. The program must keep track of the current value of DLAB and set it to the appropriate state before it attempts to read or write one of the shared-address registers. For example, if the program wishes to write to the Interrupt Enable Register, it must first set DLAB to 0 by writing to the Line Control Register and then use address 1 to get to Interrupt Enable.

Even this does not totally solve the problem, because when DLAB is 0 we still have two registers sharing address 0. Based on the material in the previous chapter, we can observe that it makes no sense for the program to have the ability to write data into the Receive Buffer or to read data from the Transmit Hold Register. The 8250 is therefore designed such that an I/O read cycle to address 0 will always transfer the contents of the Receive Buffer to the microprocessor, and an I/O write cycle will always move data from the microprocessor to the Transmit Hold Register. These procedures are summarized in Table 4–2.

Why did the designers of the 8250 choose to implement this scheme? They could have added one more address input (A3) and then given each register a unique address. In integrated circuit manufacturing, there are many economies that are related to pin count, and designers frequently add some circuit complexity to save a pin. The 8250, however, has three chip select inputs, which is more than sufficient for most applications. In the IBM ACA, for example, only CS2 is used. Why was a chip select pin not traded for another address input? It is a mystery.

TABLE 4–2 Use of DLAB for Register Selection

DLAB STATE	ADDRESS	I/O CYCLE	REGISTER ACCESSED
0	0	READ	Receive Buffer
	0	WRITE	Transmit Hold
	1	X	Interrupt Enable
1	0	X	Divisor Latch (LS Byte)
	1	X	Divisor Latch (MS Byte)

4·2 LINE CONTROL REGISTER

The Line Control Register (LCR) allows the program to set all of the asynchronous format parameters described in Chapter 2. These settings will affect both the transmit and receive functions of the 8250. The program can also check the currrent settings by reading this register. The bit layout of the LCR is shown in Table 4–3.

The first field (W) determines the number of data bits sent in transmitted characters and expected in received characters. The encoding is straightforward. See Table 4–4.

The next field (S) determines the minimum stop interval used for character transmission. Things get a little more complicated here. A single bit can encode only two values, but there are three commonly used values for the minimum stop interval: 1, 1.5, and 2 times the bit time. It was observed, however, that 1.5 is only used in equipment that operates with 5 data bits (these devices are the direct descendents of the Baudot distributor). The value represented by the 1

TABLE 4–3 Line Control Register

7	6	5	4	3	2	1	0
D	B	P			S	W	

W: Word Length (Number of Data Bits)
S: Minimum Stop Interval
P: Parity Specifier
B: Generate Break
D: DLAB

TABLE 4–4 W-Field

NUMBER OF DATA BITS	LCR BIT 1	LCR BIT 0
5	0	0
6	0	1
7	1	0
8	1	1

TABLE 4–5   S-Field

NUMBER OF DATA BITS	LCR BIT 2 (S)	MIN STOP INTERVAL
5	0	1 * T
5	1	1.5 * T
6, 7 ,8	0	1 * T
6, 7 ,8	1	2 * T

state of the S field was therefore redefined for the case where we have a 5-bit W-field. See Table 4–5.

The 8250 uses the 3-bit P-field to specify the parity rule that will be used for transmission and checked on received characters. Five different varieties of parity are available. In addition to the ODD, EVEN, and NONE options previously discussed, we can have MARK or SPACE parity. In these latter cases the parity bit is inserted following the last data bit, but it is unconditionally forced to the selected state regardless of the number of 1s in the data field. Note that MARK parity effectively adds one more bit time to the minimum stop interval.

The P bits are defined as

LCR 3:   Parity enable
LCR 4:   Parity select
LCR 5:   Stick parity

LCR 5 turns on the unconditional MARK and SPACE options in which the parity bit is effectively "stuck." The code combinations for parity are given in Table 4–6.

The B field allows the 8250 to generate a line break. This cannot be done by normal character transmission since loading 00 (hex) into the transmitter will

TABLE 4–6   S-Field

LCR BIT:	3 ENABLE	4 SELECT	5 STICK	PARITY TYPE
	0	X	X	none
	1	0	0	odd
	1	1	0	even
	1	0	1	mark
	1	1	1	space

not produce a spacing condition greater than a full character time. Only the data bits of the character are loaded into the Transmit Hold Register, and, at the beginning of the stop interval, the 8250 returns the line to the MARK state. If the B bit in the LCR is set to "one," the outgoing line is forced to SPACE and remains there until B is cleared. The timing is left to the program.

D (LCR bit 7) represents DLAB, which we have already described.

Let us look at an example of the use of the Line Control Register. What hexadecimal pattern should be loaded into LCR to produce the following data character format?

- 5 data bits
- no parity
- 1.5-bit stop interval

For 5 data bits select W = 00 (see Table 4–4). For no parity select P = 000 (see Table 4–6). For a 1.5-bit stop interval (with 5 data bits), select S = 1 (see Table 4–5). Assume that the break generator (B) and DLAB bits are 0. To put all of this together, place the previous selections in the LCR bit locations indicated by Table 4–3:

D	B	P	S	W
0	0	000	1	00

The equivalent hexadecimal value is therefore 04.

In Section 4•9 we will see another example of LCR encoding.

4·3 BAUD RATE GENERATOR

The 8250 allows the operating baud rate of the asynchronous line to be adjusted by the microprocessor program. This mechanism is shown in Figure 4–2. The 1.8432-MHz crystal oscillator input is applied to a *variable frequency divider*. This is simply a counter whose division ratio (sometimes called the *modulus* of the counter) is the binary number stored in the *Divisor Latch*. This is a 16-bit number, and the Divisor Latch must therefore be loaded in two steps. We set the Divisor Latch such that the output of the variable frequency divider is as close as possible to 16 * B_P_S for our desired line speed. 1.8432 MHz was chosen as the clock frequency because it is an exact multiple of many of the baud rates commonly used in communications equipment. For example, 1 843 200/12 is equal to 16 * 9600. Table 4–7 lists all the commonly used baud rates and the hexadecimal values required in the Divisor Latches. The decimal equivalents are also listed. Which standard baud rates cannot be exactly derived from 1.8432 MHz?

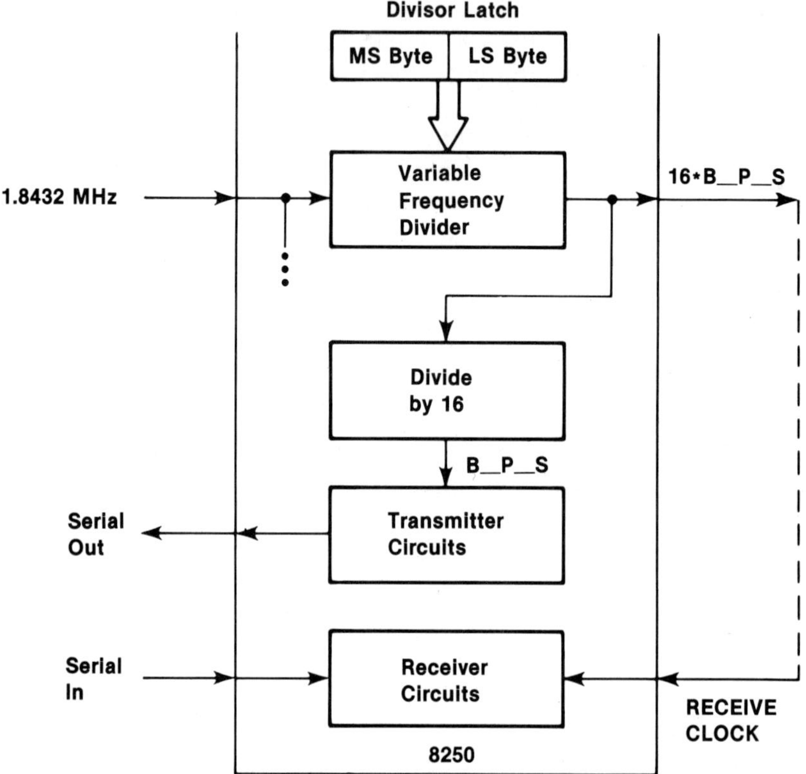

FIGURE 4–2   Baud Rate Generation

TABLE 4–7   Divisor Latch Settings

| | DIVISOR LATCH (HEX) | | DIVISOR LATCH |
LINE SPEED	MS Byte	LS Byte	(DECIMAL)
50	09	00	2304
75	06	00	1536
110	04	17	1047
134.5	03	59	857
150	03	00	768
300	01	80	384
600	00	C0	192
1200	00	40	96
1800	00	3A	64
2000	00	30	58
2400	00	20	48
3600	00	18	32
4800	00	10	24
9600	00	0C	12

The 16 * B_P_S output of the baud rate generator is not applied directly to the 8250 receive circuits. It is brought to an output pin where it is usually connected by a jumper to the RECEIVE CLOCK input. We have such a connection on the IBM ACA. We could, however, use a separate external source for the RECEIVE CLOCK. The variable frequency divider output is also reduced to B_P_S and applied directly to the transmit circuit. The external connection for RECEIVE CLOCK gives the user the option of running the 8250 transmitter and receiver at different baud rates.

4·4   INTERRUPT ENABLE REGISTER

There are four categories of events that can cause interrupts from the 8250. The *Interrupt Enable Register* allows the program to individually enable or disable each category by assigning a bit to each one. If a bit is set to the 1 state, its respective interrupt is enabled. The register layout is shown in Table 4–8.

TABLE 4–8   Interrupt Enable Register

7	6	5	4	3	2	1	0
	------------------			MS	LS	T	R

R: Enable Receive Data Available Interrupt
T: Enable Transmit Hold Register Empty Interrupt
LS: Enable Receive Line Status Interrupt
MS: Enable Modem Status Interrupt

The first two categories are associated with the assertion of the two basic data flags that we discussed in Chapter 3. The line status interrupt is triggered by any of the following events:

- overrun error
- parity error
- framing error
- receive line break

There are bits in other registers which allow the program to determine exactly which type of event caused the interrupt. Modem status interrupts are related to a subject that is covered in a later chapter and will not be discussed here.

As an example of Interrupt Enable Register encoding, what is the significance of the pattern 03 (hex) in this register?

Expanding 03 (hex) to binary and matching the bit pattern to the fields defined in Table 4–8, we have

				MS	LS	T	R
0	0	0	0	0	0	1	1

We therefore have only receive-data-available and transmit hold-register-empty interrupts enabled.

In Section 4•9 we will see another example of interrupt enable register encoding.

4•5 INTERRUPT IDENTIFICATION REGISTER

There is only one interrupt output pin on the 8250 and it is asserted on the occurrence of any of the four categories previously described. As a result the microprocessor's interrupt service routine for the 8250 must read the *Interrupt Identification Register* immediately after each interrupt is detected to determine the cause. Each of the four interrupt categories is assigned a *priority*. If several interrupts are pending simultaneously, the Interrupt Identification Register reflects the one with the highest priority. Lower priority interrupts are not lost. After the program recognizes and clears the first category, the interrupt pin is again asserted, and the Interrupt Identification Register now reflects the next highest priority pending event. The register layout is shown in Table 4–9.

TABLE 4–9 Interrupt Identification Register

7	6	5	4	3	2	1	0
-------------------------					ID		P

The P bit is 0 if any interrupt is pending, otherwise it is 1. The ID field is encoded (in decreasing order of priority) as follows:

INTERRUPT CATEGORY	ID FIELD	
	Bit 2	Bit 1
Receive Line Status	1	1
Receive Data Available	1	0
Transmit Hold Empty	0	1
Modem Status	0	0

4·6 RECEIVE BUFFER AND TRANSMIT HOLD REGISTER

As we have previously described, these are the 8-bit registers that provide the direct link between the program and the communications line. Only the character's data bits appear in these registers. The start bit, the parity bit, and the stop interval are added by the 8250 on transmission and removed by the 8250 for received characters. If the 8250 is configured for less than 8 data bits, the character appears right justified in these registers.

4·7 LINE STATUS REGISTER

The *Line Status Register* contains a number of flags and indicators, most of which are associated with interrupt events. The bit layout is shown in Table 4–10.

DR is the primary indicator for the receive function. OE, PE, FE, and BI are all associated with line status interrupts and can be used by the program to determine the exact cause of the interrupt. HE is the primary indicator for the transmit function.

In Chapter 3 we did not show a program-readable flag for the state of the Transmit Shift Register. The 8250 does provide this feature, but it is not associated with any interrupt. By reading the state of SE, the program can determine when the last character of an outgoing message has been completely shifted onto the line.

There are two additional registers in the 8250: *modem control* and *modem status*. Since we will not discuss modems until a later chapter, these registers will not be described here.

TABLE 4-10 Line Status Register

7	6	5	4	3	2	1	0
-----	SE	HE	BI	FE	PE	OE	DR

DR: Data Ready
OE: Overrun Error
PE: Parity Error
FE: Framing Error
BI: Break Interrupt
HE: Transmit Hold Register Empty
SE: Transmit Shift Register Empty

4·8   THE ASYNCHRONOUS COMMUNICATIONS ADAPTER

Although the UART is the heart of a small computer's communications interface function, it is not sufficient by itself. At the very least, the following functions must be added:

PACKAGING   Since asynchronous communications is frequently an option in personal computer systems, the UART and its auxiliary circuits must be packaged in a form that is easily handled, inserted, and removed. The package is usually a small printed circuit board.

BUS INTERFACE   Since address and data bus signals vary somewhat from one microprocessor to another, we must include circuits that make these bus signals compatible with the format expected by the UART.

COMMUNICATIONS LINE INTERFACE   The *Serial In/Out* pins of the UART may need voltage or current level conversion to interface with a modem or locally connected terminal. This conversion does not change the digital nature of the signals. We have not performed any modulation yet.

In the IBM PC all of these functions are provided by a small printed circuit board called the *Asynchronous Communications Adapter* (ACA). The major

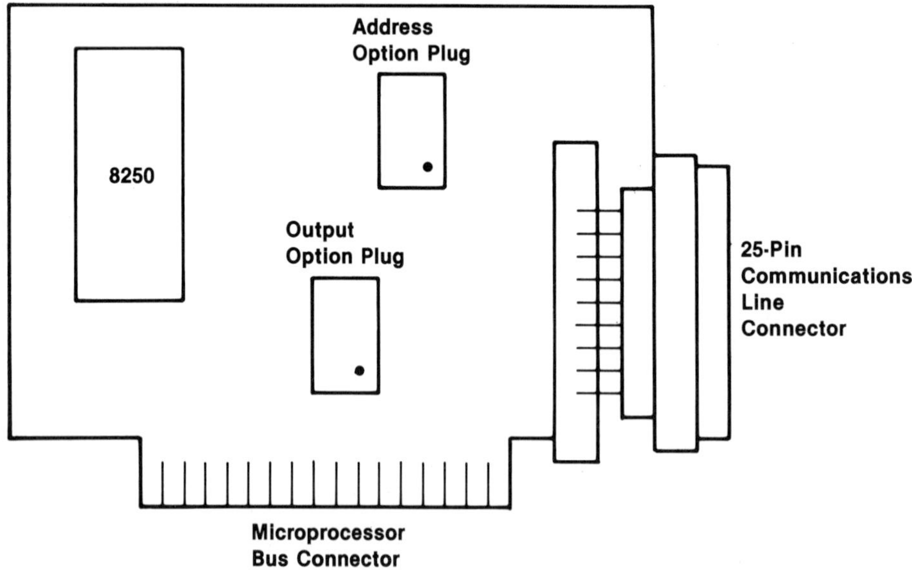

FIGURE 4–3   IBM Asynchronous Communications Adapter

components of this module are illustrated in Figure 4–3. In addition to the 8250 and some auxiliary chips, the ACA contains two *option plugs* which allow the user to select features that will be described. Each option plug is marked with a dot. The two possible states of each option are selected by inserting the plugs either with the dot up or with the dot down. The microprocessor bus interface is a 62-pin printed circuit edge-connector, and the communications line interface is the standard 25-pin cable-connector used on terminals and modems.

Figure 4–4 is a block diagram of the ACA. The microprocessor interface, at the left side of the drawing, is composed of the address bus, the data bus, the interrupt line, and a few strobes that are not shown.

The address bus contains ten wires (A9–A0). As described in Chapter 3, the bus is split into a least significant section (A2–A0), which is used for 8250 register selection, and a most significiant section (A9–A3), which is used to form chip select. One of the option plugs also affects the chip select decode. This operates as follows:

- The address decoder always expects A9 and A7–A3 to be in the 1 state for chip select.
- If the address option plug is inserted with the dot up (*primary* position), the address decoder expects A8 to be 1 for chip select.
- If the address option plug is inserted with the dot down (*alternate* position), the address decoder expects A8 to be 0 for chip select.

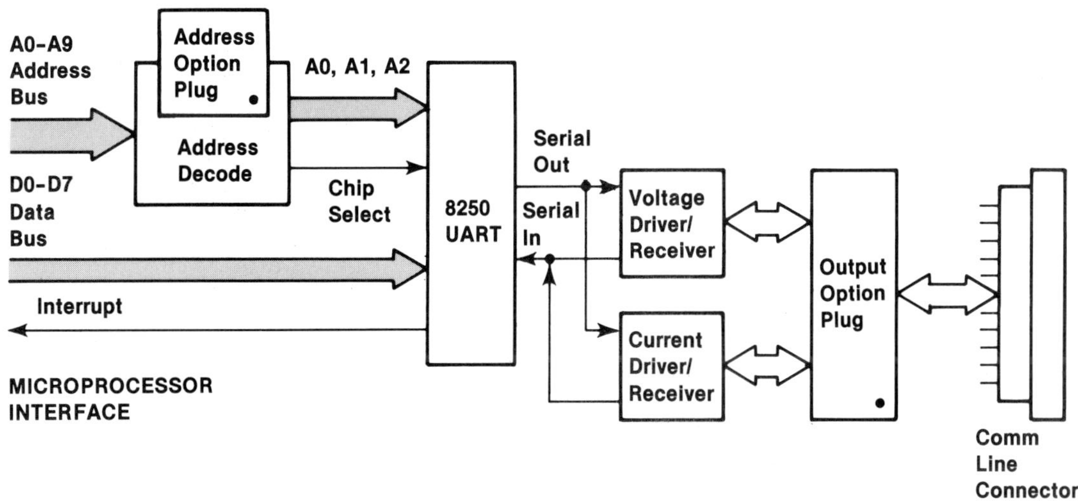

FIGURE 4–4 IBM ACA Block Diagram

Each register, therefore, has two different I/O addresses, depending on the position of the address option plug:

	A9	A8	A7	A6	A5	A4	A3	A2	A1	A0
Primary:	1	1	1	1	1	1	1	(	Reg Sel	)
Alternate:	1	0	1	1	1	1	1	(	Reg Sel	)

Chip Select

For example, the addresses of the line control register (A2–A0 = 0 1 1) would be

Primary:	1 1	1 1 1 1	1 0 1 1	[= 3FB (hex)]
Alternate:	1 0	1 1 1 1	1 0 1 1	[= 2FB (hex)]

All other registers will also have two different addresses which are most frequently referenced by their 3-digit hexadecimal values. For the primary option the first two digits will always be 3F (hex); for the alternate option the first two digits will always be 2F (hex). The third digit will have 1 as its most significant bit, with the remaining three bits coming from Table 4–1. The computation of the hexadecimal addresses for all the 8250 registers is left as an exercise for the student.

What is the purpose of the address option plug? It allows two ACA modules to be simultaneously plugged into the PC with a unique set of addresses for the registers of each one. As a result, the program can simultaneously manage two communications lines with an independent data format for each one. This is similar to the way memory is expanded by adding multiple chips of the same type.

The communications line interface of the ACA is shown at the right side of Figure 4–4. The *output option plug* allows selection of two different styles of digital interface to the communications line connector:

Current interface
A current loop similar to that described in connection with the Baudot distributor. The loop operates at approximately 20 mA.

Voltage interface
A signal interface driven at voltage levels compatible with the RS-232C specification that will be described in a later chapter.

All modems require the voltage interface. Older terminals, such as electromechanical teletypes, use the current interface. Newer terminals, such as the familiar CRT devices, generally use the voltage interface but frequently offer current loop as an option. In the IBM PC, the wires for both interface types are routed to a common connector but use different groups of pins.

4·9   A TYPICAL COMMUNICATIONS PROGRAM

We are now ready to put all of this material together and examine the details of a typical program that would employ the services of a UART. To avoid involvement with the intricacies of programming languages, we will not actually write the program, but its anatomy will be described in detail.

Suppose we have a single-user personal computer with the hardware illustrated in Figure 4–5. This computer is used by the administrator of a school. The disk holds a *data base,* which is a large collection of related pieces of information. In the case of a school the data base might include records concerning classes, teachers, students, grades, and so on. From time to time, the administrator will formulate inquiries to the data base. For example: "List the names of all seniors in Mr. Smith's American history class who had failing grades at midterm." The question would have to be formulated in terms that looked a lot less like normal English than the example quoted, but some number of characters would have to be keyed-in on the administrator's terminal, transmitted via the communications line, and sensed by the microprocessor's program. The program would determine the answer and return it to the terminal via the communications line. This style of computer application is called *interactive processing* and is common throughout modern data processing installations.

As far as the communications adapter is concerned, the program has two phases:

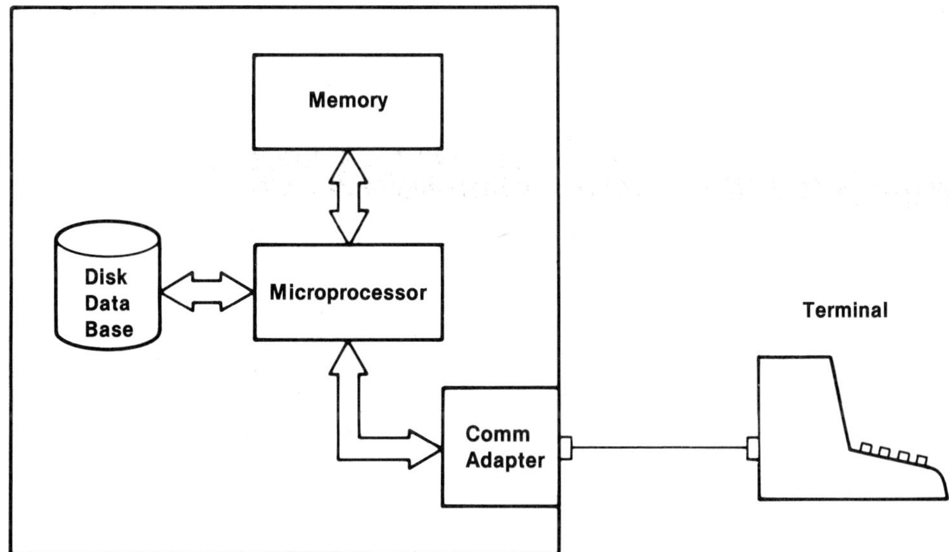

FIGURE 4–5   Typical Application Hardware

Initialization
This phase runs once, at the time the program starts. It establishes the data format parameters and leaves the 8250 ready for transaction processing.

Transaction processing
After initialization this phase of the program accepts inquiries from the terminal, finds answers from the data base, returns this response to the terminal, and waits for the next inquiry.

We will describe the interaction between the program and the 8250 for each phase.

Initialization

During initialization we establish the basic operating parameters that were decided upon by the designers of the program. Suppose that our administrator's terminal is an inflexible device that will only accept the following data format:

- 7 data bits
- odd parity
- 2-bit stop interval
- 134.5-baud data rate

Since this is a single-user system, the designer has decided not to use interrupts for transmitting and receiving characters from the 8250. Also, since the terminal in Figure 4–5 is locally connected, we do not have to bother with modem status interrupts (there is no modem present). We are, however, concerned with error handling, but, since errors occur infrequently, we do not want to bother checking for them on each received character. We can accomplish this by putting error handling in an interrupt service routine. This will require that the line status interrupt be enabled.

Initialization is accomplished by a series of OUT instructions, each of which will specify an 8250 register and the 8-bit data pattern that will be loaded into the register. Each step in Figure 4–6 represents one of these instructions.

STEP	REGISTER	ADDRESS (HEX)	DATA BINARY	HEX	COMMENT
1	Line Control	3FB	1XXX XXXX	80–FF	Set DLAB
2	Divisor Latch (MS Byte)	3F9	0000 0011	03	134.5 Baud
3	Divisor Latch (LS Byte)	3F8	0101 1001	59	134.5 Baud
4	Line Control	3FB	0000 1110	0E	Clear DLAB, Set Format
5	Interrupt Enable	3F9	0000 0100	04	Line Status Only

FIGURE 4–6 Asynchronous Communications Adapter Initialization

Addresses are specified as if we were using an IBM PC ACA with the primary address option selected.

Since the baud rate will be set once and never changed, we may as well get it done immediately. We check Table 4–7 and determine that 134.5 baud requires 0359 (hex) in the Divisor Latches. Before we can load this, we must set DLAB. This is done in Figure 4–6, Step 1. At this point, none of the format parameters have been set, so we do not really care what goes into the Line Control Register bits other than DLAB. If we wanted to change the value of DLAB at some point after data format parameters had been set, we would have had to be careful about not changing other bits in the Line Control Register.

In Steps 2 and 3 we load the Divisor Latches with the appropriate number for 134.5 baud. We are now finished with the Divisor Latches and can safely clear DLAB. This requires writing to the Line Control Register again, and—since this register also contains the fields for setting stop interval, parity, and the number of data bits—we can combine several operations into one instruction. This is shown in Step 4. Such combined operations are a good example of why computer programs become difficult to read, and the student should verify that the bit pattern loaded in Step 4 does, in fact, produce the desired format.

The only remaining requirement is to enable the line status interrupt, which is the only one we want. This is done in Step 5. Note that the addresses in Steps 2 and 5 are both 3F9 (hex). If we had not cleared DLAB at some point between Steps 2 and 5, the data in Step 5 would have gone into the wrong register. Rather than enabling the line status interrupt, Step 5 would have changed the baud rate to some mysterious value. The programmer would probably check the values that were loaded in Steps 2 and 3, conclude that they are correct, and blame the problem on the hardware.

Transaction Processing

The 8250 in the ACA is now ready to deliver data characters to the transaction processing section of our program. Inquiries formulated by the terminal operator will be strings of text, typically a few dozen characters in length. The operator is usually required to indicate the end of the inquiry message by typing a predefined *message terminator* key such as *carriage return* or *enter*. The operator will then wait for the program's response to appear on the screen.

The transaction processing program generally cannot analyze the operator's inquiry until the entire message has been received. The program will therefore set aside (the buzzword is *allocate*) a block of memory bytes large enough to hold the biggest possible valid inquiry. This memory block is called a *buffer*. Since this is a single-user program, there is now nothing to do but wait for incoming data from the terminal. The program will repeatedly test the 8250 DR flag. Since the program is orders of magnitude faster than the quickest typist, most of the time nothing will be present. When a character does arrive, it will be stored in the first available byte in the memory buffer and tested to see if it matches the message terminator character. If not, the program will wait for more incoming data, storing each character in successive buffer locations. Finally, the operator will type the terminator character. The program can now analyze the full inquiry in the buffer to formulate the response message. In the case of our data base application this will take at least one and probably several read operations to the disk drive. A second memory buffer must be allocated to hold the data read from the disk. The data from the disk may have to be reformatted to put it into a form which can be more easily understood by the operator, and this will require a third memory buffer. The response is now ready to go. The first two buffers may now be released to a *free memory pool,* and the third buffer, containing the response message, is handed over to the communications transmitter section of our program.

The transmitter program places the first character from the response buffer into the 8250 Transmit Hold Register and then continously monitors the hold-register-empty flag. Each time the flag becomes true, another character will be transmitted until the entire response message has been sent. The operator sees the response on the CRT screen and, if desired, may formulate another inquiry. The program is now prepared to process the next transaction.

The point that must be emphasized in this example is that, even in small computers, many single-user programs spend almost all of their time waiting for manually typed input, for the disk drive to mechanically move its read head to the track where the desired information resides, and for the Transmit Hold Register to become empty.

REVIEW QUESTIONS

1. What is the effect of operating the IBM ACA at a baud rate that cannot be exactly derived from 1.8432 MHz by integer division?

2. List the hexadecimal IBM PC bus addresses for all the registers in the 8250 on the ACA (list both the primary and alternate option addresses).

3. Write a series of steps to initialize the 8250 on the IBM ACA (alternate address option) for

   - 8 data bits
   - SPACE parity
   - 110-baud data rate
   - 1-bit stop interval
   - all interrupts enabled except modem status

   Indicate the function of each step and specify both the register address and data content in hexadecimal.

4. Describe the series of steps a program would have to take to change the baud rate of an 8250 without disturbing the settings of any other parameters such as parity, number of data bits, or stop interval.

5. A program reads the line status register of an 8250 and finds that it contains 63 (hex). What would you conclude has just occurred within the 8250 receiver?

6. Specify the 8250 bit that you would set to transmit a 3-second spacing condition.

7. An 8250 interrupt service routine reads the Interrupt Identification Register and finds that it contains 02 (hex). What binary values would you expect to find in bits 5 and 6 of the Line Status Register at this time?

chapter five

CODES AND TERMINALS

In Chapter 2 we briefly referred to the subject of codes. We stated that a code was a set of binary numbers, each of which has been defined to represent an alphanumeric symbol or a control function for a particular piece of digital equipment. On an asynchronous communications line, the binary value of the code appears as the data bits of the transmitted character. Devices such as printers and CRT displays convert these binary values to their familiar visual representations or respond to the specified control function. In normal data communications practice, all the binary values (characters) in a given code have the same number of bits. The number of bits per character is the primary factor that determines the maximum number of symbols that can be represented by a code. If the number of bits per character is N, the maximum number of characters that we can possibly define is

$$MAX\_CHARS = 2^N$$

The available binary numbers for representing the characters will range from 0 to MAX_CHARS $-$ 1. For example, a 6-bit code can provide 64 different binary values, ranging from 0 to 63.

In the next few sections we will describe some codes commonly used in data communications equipment. As we shall see, not all codes make use of all the binary values at their disposal and, in one case, some trickery is used to make each binary value represent two different symbols. Some codes provide a parity bit and some do not. Other incompatibilities also exist, and, since most equipment is designed to operate with only one code, these inconsistencies present difficulties for designers of communications networks. There has been some attempt at standardization, but users will not discard operational equipment just because a new standard has been published. Equipment manufacturers often choose to ignore standards or to interpret them in their own unique fashion. Consequently, data communications is often hindered by "language barriers."

5·1 BAUDOT

This is the code that was transmitted by the Baudot distributor, and it is still found in many low-speed electromechanical teletypes. It uses 5 data bits; has no parity; has a 1.5-bit minimum stop interval; and operates at speeds of 50 or 75 baud. The *Telex* system is a switched network that can interconnect teletypes in much the same manner that the more familiar voice network connects telephones. The Telex network uses Baudot code at 50 baud and has been implemented in virtually every country in the world. Because of its large installed base of equipment and wide distribution, this slow and antiquated system continues to operate in the midst of all our high-tech electronic equipment.

A chart showing the bit patterns of Baudot code and their associated symbols is shown in Figure 5–1. There are actually several different 5-bit codes that are commonly called Baudot. The one illustrated in Figure 5–1 has been standardized by an agency of the United Nations as *International Telegraph Alphabet #2*. Note that most of the bit patterns are associated with two different symbols. The 5 data bits of Baudot can represent 32 different symbols, insufficient for even the alphabet and the digits 0–9. To circumvent this problem, Baudot contains two special characters called *figures shift* (11011) and *letters shift* (11111) whose behavior resembles the operation of the shift lock key on a conventional typewriter. These characters toggle the receiving device between two different states called *cases*. The receiver must remember the most recent shift character and interpret all subsequent characters on the basis of the current

LTRS	FIGS	Bit Pattern 5 4 3 2 1	LTRS	FIGS	Bit Pattern 5 4 3 2 1
A	–	0 0 0 1 1	Q	1	1 0 1 1 1
B	?	1 1 0 0 1	R	4	0 1 0 1 0
C	:	0 1 1 1 0	S	'	0 0 1 0 1
D	$	0 1 0 0 1	T	5	1 0 0 0 0
E	3	0 0 0 0 1	U	7	0 0 1 1 1
F	!	0 1 1 0 1	V	;	1 1 1 1 0
G	&	1 1 0 1 0	W	2	1 0 0 1 1
H	#	1 0 1 0 0	X	/	1 1 1 0 1
I	8	0 0 1 1 0	Y	6	1 0 1 0 1
J	Bell	0 1 0 1 1	Z	"	1 0 0 0 1
K	(	0 1 1 1 1	Letters Shift		1 1 1 1 1
L	)	1 0 0 1 0	Figures Shift		1 1 0 1 1
M	.	1 1 1 0 0	Space (SP)		0 0 1 0 0
N	,	0 1 1 0 0	Carriage Return		0 1 0 0 0
O	9	1 1 0 0 0	Line Feed		0 0 0 1 0
P	0	1 0 1 1 0	Blank		0 0 0 0 0

FIGURE 5–1 Baudot Code

case. Shift characters themselves do not print. For example, the pattern 00011 represents both A and − . If a Baudot teletype received the sequence

LTRS   FIGS

(11111)   (00011)   (00011)   (11011)   (00011)

the characters AA − would be printed. The same is true for B and ?, C and :, and so forth. This is cumbersome, but it provides the full alphabet (uppercase only), the digits, and the most common punctuation marks. There are a few characters (carriage return, line feed, and space) which appear in both cases. Also notice the *bell* character in the figures case. This literally causes a bell to ring at the receiving teletype. In the Telex system the caller often uses this device to attract the attention of an operator at the called teletype. Baudot was a practical and useful system within the limitations of electromechanical equipment.

There are, nevertheless, a number of difficulties with Baudot code that are not present in the more modern codes. The effective transmission rate, in printed characters per second, is reduced by the necessity of inserting shift characters. Shift characters must also be manually inserted from the keyboard of a Baudot teletype, making it slow and difficult to use. Baudot does not include parity or any other automatic error detection mechanism. For this reason all numeric quantities in Telex messages were frequently repeated at the end of the message, in the hope that a human reader would notice any discrepancies. Also, if the line error happened to modify a bit in a shift character, the message might be rendered totally unintelligible. Finally, Baudot only includes printable alphanumeric characters. We will shortly see that modern terminals and communications equipment require many control functions that are not directly displayed or printed.

5·2   EBCDIC

Over a period of many years, the IBM Corporation developed a number of different codes which attempted to solve some of the deficiencies of Baudot. Among the added features were

- lowercase alphabetic characters
- elimination of figures/letters shifts
- character parity bit
- control functions

None of these codes addressed all of the problems simultaneously, and, in some cases, the control functions were highly specific to a particular variety of equipment such as paper-tape reader/punches. The end result of this evolution (in 1962) was a more powerful and general purpose code called *Extended Binary Coded Decimal Interchange Code* (EBCDIC), commonly pronounced EEBSAdik. This code is common throughout IBM's line of large computers and communications equipment, but it is not usually used in their personal computers.

The EBCDIC code is illustrated in Figure 5–2. EBCDIC characters have 8 data bits and can be represented numerically as a 2-digit hexadecimal quantity. The code is therefore presented as a matrix with a character's row indicating the most significant digit and its column indicating the least significant digit. For example, the numeric value for A is C1 (hex).

There are several observations we can make from the EBCDIC matrix. The upper four rows contain all control functions. These are similar to control functions available in other modern codes, and we will describe most of their uses later. The next four rows contain punctuation marks followed by three rows of lowercase alphabetics, a blank row, three rows of uppercase alphabetics, and a row for the numeric digits. This logical and consistent grouping of character values is advantageous for the programs that are interpreting and manipulating messages composed of EBCDIC characters. For example, we could determine if a character is a control function by merely testing whether or not its value was less than 40 (hex). We can easily convert the hexadecimal values for lowercase letters to their equivalents for uppercase, a + 40 (hex) = A. Also, the actual bi-

LEAST SIGNIFICANT DIGIT (HEX)

MOST SIGNIFICANT DIGIT (HEX)

	0	1	2	3	4	5	6	7	8	9	A	B	C	D	E	F
0	NUL	SOH	STX	ETX	PF	HT	LC	DEL			SMM	VT	FF	CR	SO	SI
1	DLE	DC_1	DC_2	DC_3	RES	NL	BS	IL	CAN	EM	CC		IFS	IGS	IRS	IUS
2	DS	SOS	FS		BYP	LF	EOB	PRE			SM			ENQ	ACK	BEL
3			SYN		PN	RS	UC	EOT					DC_4	NAK		SUB
4	SP										¢	.	<	(	+	|
5	&										!	$	*	)	;	¬
6	−	/										,	%	—	>	?
7											:	#	@	'	=	"
8		a	b	c	d	e	f	g	h	i						
9		j	k	l	m	n	o	p	q	r						
A			s	t	u	v	w	x	y	z						
B																
C		A	B	C	D	E	F	G	H	I						
D		J	K	L	M	N	O	P	Q	R						
E			S	T	U	V	W	X	Y	Z						
F	0	1	2	3	4	5	6	7	8	9						

FIGURE 5–2   EBCDIC Code Matrix

nary value of a numeric digit is easily derived from its EBCDIC code. For example, 0 = F0 (hex); 1 = F1 (hex); 2 = F2 (hex); etc. All of these advantages over Baudot stem from the fact that EBCDIC was designed expressly for the environment of electronic computers rather than electromechanical teletypes.

There are still some weaknesses in EBCDIC. The first is the lack of a parity bit. This is not a major flaw because many of the communications devices that use EBCDIC also incorporate error detection techniques that are much more powerful than character parity. (We will study these techniques in a later chapter.)

The second weakness is more serious and quite obvious from Figure 5–2. Almost half of the locations in the matrix are empty. The 8 data bits in EBCDIC provide 256 different binary values. Only 139 of these are actually assigned to symbols or control functions. If the designers of EBCDIC could have agreed to reduce the character set from 139 to 128 (perhaps by eliminating some control codes or punctuation marks), the number of data bits could have been correspondingly reduced from 8 to 7. This alone would reduce the transmission time for any message coded in EBCDIC by approximately 12%. Since telephone charges are often involved, this reduction would be very beneficial for users.

There is, however, another way of looking at the empty space in the EBCDIC matrix. Suppose we developed a new type of equipment whose needs could not be satisfied by the existing EBCDIC character set. This new device may have to transmit symbols from alphabets other than English (German umlauts, for example) or may require control functions that are not presently defined. This flexibility could be provided by adding definitions for the empty positions in the matrix. We will see in the next few sections, however, that flexibility and expansion are possible without leaving holes in our code matrix.

5·3   ASCII

The *American National Standard Code for Information Interchange* was introduced in 1963, and the current standard was published in 1967. The most commonly used acronym is ASCII, which is pronounced AS-key. The code was specifically designed for electronic communications and data processing equipment and is probably the most widely used information code in the world today.

The code matrix for ASCII is illustrated in Figure 5–3. ASCII is a 7-bit code. It can therefore be represented by two hexadecimal digits, but, unlike EBCDIC, the most significant digit will never exceed 7. In Figure 5–3 a character's most significant digit is indicated by its column and its least significant digit by its row. For example, the value for A is 41 (hex). An optional parity bit can be added to ASCII characters, and many communications devices use this feature. The choice of odd or even parity is left to the equipment designer.

As was the case in EBCDIC, the ASCII matrix is logically organized. Columns 0–1 contain control characters. Their names and functional classifications

**MOST SIGNIFICANT
DIGIT (HEX)**

	0	1	2	3	4	5	6	7
0	NUL	DLE	SP	0	@	P	`	p
1	SOH	DC1	!	1	A	Q	a	q
2	STX	DC2	"	2	B	R	b	r
3	ETX	DC3	#	3	C	S	c	s
4	EOT	DC4	$	4	D	T	d	t
5	ENQ	NAK	%	5	E	U	e	u
6	ACK	SYN	&	6	F	V	f	v
7	BEL	ETB	'	7	G	W	g	w
8	BS	CAN	(	8	H	X	h	x
9	HT	EM	)	9	I	Y	i	y
A	LF	SUB	*	:	J	Z	j	z
B	VT	ESC	+	;	K	[	k	{
C	FF	FS	,	<	L	\	l	|
D	CR	GS	–	=	M	]	m	}
E	SO	RS	.	>	N	^	n	~
F	SI	US	/	?	O	_	o	DEL

LEAST SIGNIFICANT DIGIT (HEX)

FIGURE 5–3   ASCII Code Matrix

are listed in Figure 5–4. Since most of their functions are related to the operation of terminals and communications protocols, we will postpone discussion of the individual control characters until we get to these subjects. Columns 2–3 contain numerics and punctuation marks. Columns 4–5 contain uppercase alphabetics, and columns 6–7 contain lowercase alphabetics. There are a few miscellaneous punctuation marks scattered through columns 4–7. There are no empty spaces in the matrix.

All of the convenience features for programmers are present:

- all control characters have values less than 20 (hex);
- a numeric digit's binary value is equal to the least significant digit of its ASCII code, 0 = 30 (hex), 1 = 31 (hex), etc.
- conversion between uppercase and lowercase alphabetics is easy, A + 20 (hex) = a.

The relationship of control characters to the alphabetics is interesting because this relationship is usually used on asynchronous terminals to provide the operator with the ability to transmit any desired control character. Most

Char	Name	Fill	Information Separator	Format Effector	Comm Control
NUL	Null	X			
SOH	Start of Header		X		
STX	Start of Text		X		
ETX	End of Text		X		
EOT	End of Transmission		X		
ENQ	Enquiry				X
ACK	Acknowledge				X
BEL	Bell				X
BS	Backspace			X	
HT	Horizontal Tab			X	
LF	Line Feed			X	
VT	Vertical Tab			X	
FF	Form Feed			X	
CR	Carriage Return			X	
SO	Shift Out		X		
SI	Shift In		X		
DLE	Data Link Escape				X
DC1	Device Control 1				X
DC2	Device Control 2				X
DC3	Device Control 3				X
DC4	Device Control 4				X
NAK	Negative Acknowledge				X
SYN	Synchronous Idle	X			
ETB	End of Transmission Block		X		
CAN	Cancel		X		
EM	End of Medium		X		
SUB	Substitute		X		
ESC	Escape		X		
FS	File Separator		X		
GS	Group Separator		X		
RS	Record Separator		X		
US	Unit Separator		X		
DEL	Delete	X			

FIGURE 5–4   ASCII Control Characters

ASCII terminals and personal computers do not have any explicit keys for the control characters. They do, however, provide a "control key" which is usually labeled CTL or CTRL. The CTRL key is used in a manner similar to a shift key. That is, it is held down while a second key is struck. The operation is as follows. If the CTRL key is held down and we hit a second key that is labeled with a character from columns 4 or 5 of the ASCII matrix, the value actually transmitted by the terminal is 40 (hex) less than the ASCII value normally represented by the key that was hit. We can express this as

$$CTRL\_KEY = KEY\_VALUE - 40 \text{ (hex)}$$

For example, what is transmitted if CTRL is held down and we strike the key labeled "A"?

$$CTRL\_A = A - 40 = 41 - 40 = 01 \text{ (hex)}$$

This is exactly how one would transmit the control character SOH [01 (hex)].

Suppose we wanted to transmit the control character DC3. What key would we hit while CTRL was held down?

$$DC3 = 13 \text{ (hex)} = KEY\_VALUE - 40$$
$$KEY\_VALUE = 13 + 40 = 53 \text{ (hex)} = S$$

What the control key is actually doing is translating a key value four columns to the left on the ASCII chart. Symbols in column 4 go to column 0; symbols in column 5 go to column 1. Also, note that what matters is the label on the key, not its current relation with the *shift* or *shift lock* keys. Keys are always labeled with the uppercase representation of their associated alphabetic symbol. Hitting the key "A" with CTRL held down will always produce the control character SOH. It does not matter whether the key would produce "A" or "a" if it was hit without CTRL held down. This even holds true for nonalphabetic keys. For example, to transmit ESC, hold down CTRL and hit "[".

5·4   TERMINALS

In Chapter 1 we defined a terminal as a data input/output device physically separated from a host computer and connected by a communications line to a port on the host computer. There is a wide variety of devices that fits this description. For the remainder of this book, however, when we speak of a terminal we shall mean the familiar device which consists of a keyboard and a display. The display may be either a CRT or a hard-copy printing device.

There are a few basic concepts commonly employed in designing terminals which strongly influence their operational characteristics. We will describe these concepts and their major features in the next two sections. With these ideas in mind, we go on to examine the manner in which the ASCII control characters are used to implement these features. We will not describe the features of specific devices, but will illustrate general principles that apply to many actual terminals.

5·5   CHARACTER MODE TERMINALS

Figure 5–5 is a block diagram of a *character mode* terminal. The terminal is comprised of two independent sections (the keyboard and the display) and is therefore a full-duplex device. The communications link is asynchronous, and

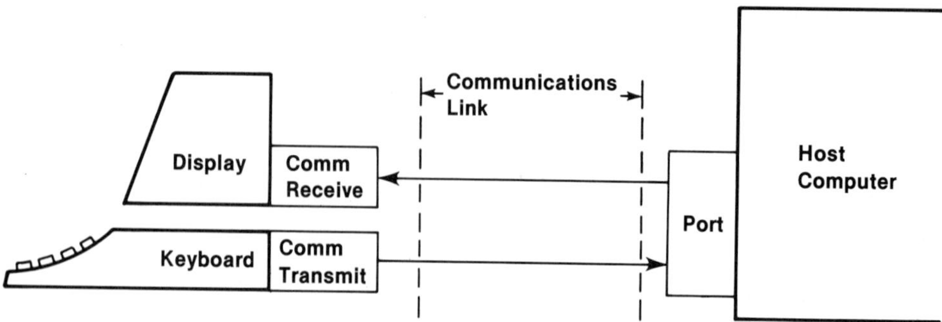

FIGURE 5–5   Character Mode Terminal Block Diagram

a character is produced for transmission to the host computer each time a key is struck. Conversely, any character transmitted from the host computer port will appear on the display. Note that there is no direct connection between the keyboard and the display. How does a character appear on the screen when a key is struck? In most cases the communications link from a character mode terminal to the host computer port is full-duplex. As soon as the program in the host computer detects the keyed character, it will retransmit the character back onto the communications link to the terminal's display device. The character will then appear on the screen or printer. This process is called *remote echo*. Remote echo provides visual confirmation of the fact that the host computer correctly received what was sent, and the speed of transmission is such that (even when modems and phone lines are involved) the operator is unaware of any delay between the keystroke and the appearance of the character. As we noted previously, computers are faster than communications lines; communications lines, however, are faster than people.

In most cases, the host-computer program will echo the same character that was transmitted from the keyboard, but it is under no obligation to do this. We will soon see some examples of situations where the host-computer program will echo something other than what was received from the keyboard.

Occasionally, a character mode terminal is connected to a host computer by a half-duplex link. As we shall see in a later chapter, the direction of transmission of a half-duplex circuit cannot be reversed (turned around) instantly. If we attempt to use our remote echo technique in this situation, the half-duplex line would have to be turned around for each keystroke, reducing the character transmission rate to a point where even the slowest typist would become impatient. This is clearly impractical. To deal with this situation, most terminals have an optional mode in which an internal connection between the keyboard and the display is enabled, thus making the character appear on the screen simultaneously with its transmission to the host computer. This is called *local echo*, and, in this mode, what appears on the screen is identical to what is typed on the keyboard.

An Asynchronous Character Mode Terminal
Courtesy of Digital Equipment Corporation

Sometimes when a new terminal is connected to a host computer port, the operator will notice that each time a key is struck, the associated character appears on the display twice. This is probably because the terminal is set for local echo and the host computer is programmed for remote echo. On many terminals, the terms used for local echo and remote echo are, respectively, half-duplex and full-duplex. In addition to local and remote echo, many terminals have a third option called *off-line* operation. In this case, the connection between the keyboard and the display is enabled, but the communications line interface is not operational. This allows the terminal to be used as a typewriter.

5·6 BLOCK MODE TERMINALS

The essential difference between *block mode* terminals and character mode terminals is that block mode devices do not transmit a character every time the operator strikes a key. Block mode terminals always use a CRT display (although they may also have an auxiliary printer), and the keystrokes appear on the screen as the operator types. The operator may review what has been typed and is able to make corrections and changes anywhere on the screen. Up to this point, not a single character has been transmitted to the host computer. When the operator is satisfied with the contents of the screen, the operator may hit a key marked ENTER or TRANSMIT and the terminal will send everything the operator typed since the previous execution of the ENTER function.

The actual process of transmitting the contents of a block mode terminal's screen requires operational complexities that were not present in character mode terminals. The message block is usually transmitted in synchronous format, and the absence of start bits requires the insertion of synchronization characters by the terminal. Block mode terminals are commonly used on multidrop circuits and cannot indiscriminately start transmission because one of the other stations may be currently active. These terminals must therefore be designed to follow the rules of a communications protocol. We will study these topics in depth later. The point being made here is that block mode terminals are more complex and therefore more expensive than character mode terminals. The compensating advantage is a reduction in the host-computer's work load because it does not have to respond to every keystroke. We have previously stated that an individual keystroke consumes only a small fraction of a host-computer's processing power, but many commercial installations have hundreds of terminals connected to a host computer. Their combined effect can be very noticeable.

To clarify some of these ideas, let us look at a typical application of a block mode terminal. Suppose we want to enter information, concerning a new student, into the school administration data base example used at the end of the previous chapter. The process we describe is called *data entry,* and it starts when a program in the host computer sends a block of text to our block mode terminal. This makes the screen appear as shown in Figure 5–6. The purpose of this screen is to make the data entry process simple to understand and do.

The operator is expected to follow a "fill-in-the-blanks" procedure. Labels such as "name:," "city:," and so on, indicate what is to go into each "blank." The operator is prevented, by the circuits in the terminal, from typing over these labels, which are therefore called *protected fields*. The areas where data can be entered are obviously *unprotected fields*. When the screen was initially transmitted from the host computer, this message block contained instructions indicating exactly which areas were to be protected and which were not. In our sample screen we have made the size of the unprotected fields explicit by displaying them in *inverse video*. If *normal video* for our terminal means that a

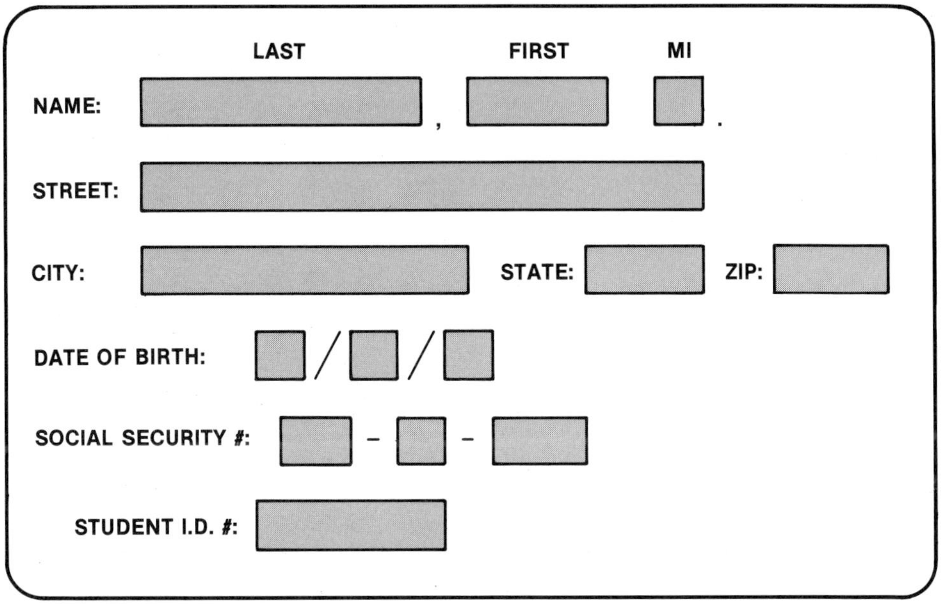

LAST   FIRST   MI

NAME:

STREET:

CITY: STATE: ZIP:

DATE OF BIRTH: / /

SOCIAL SECURITY #: - -

STUDENT I.D. #:

FIGURE 5–6   Data Entry Screen on a Block Mode Terminal

character appears as a bright image against a dark background, an inverse video character appears as a dark image against a bright background. Since nothing has yet been typed on our screen, the inverse video fields appear as brightened but empty areas. Inverse video is an example of what terminal manufacturers call a *video attribute*. If we assume that a "normal" character is displayed in normal video and at a standard intensity (brightness), some of the typical variations that are available as video attributes are

- inverse video
- low intensity
- high intensity
- underlined characters
- blinking characters

Any area of the screen has both a video attribute and a protection state. As was the case with protection, the message block from the host computer can define the video attribute of any field on the screen. In Figure 5–6, for example, we may have low intensity protected fields and inverse video unprotected fields.

We now have to enter some data into the unprotected fields. The screen will have a *cursor* (typically a rapidly blinking square, approximately the size of one character) which indicates where the next typed character will appear on the screen. As characters are typed, the cursor will advance to the right in the unprotected field. We may also have a group of *function keys* (perhaps eight or sixteen of them, with keytop labels such as F1, F2, . . .) which allow the operator to do things such as

- move the cursor within the unprotected field, allowing previously typed characters to be overwritten;
- jump the cursor from its current unprotected field to the next or previous one;
- edit the contents of unprotected fields by inserting or deleting characters in the middle of previously typed material. (The existing characters have their positions shifted as required to compensate for insertions and deletions.)

Block mode features such as these give the operator a convenient way of formatting a large quantity of information without using a single microsecond of host-computer time. If we attempted to do the same thing with a character mode terminal, the program in the host computer would have to keep track of every cursor movement and text correction entered by the operator.

If the operator is now satisfied with the appearance of the screen, the operator can hit ENTER. Typical block mode terminals may be configured to transmit the contents of the entire screen or the modified unprotected fields only. Field separator characters in the transmitted block may also vary from one terminal model to another. The program in the host computer, of course, must be written to handle exactly the format the terminal will send.

It is important to observe that the procedure we have just described requires a CRT display. The cursor motion and editing features that make this type of data entry so "user friendly" would be impossible on a hard-copy (printing) device. Also, the application we chose to illustrate used protected fields to restrict the range of cursor movement. In many applications the entire screen is one unprotected field, allowing the cursor to be moved freely and characters to be typed at any location. This allows terminals to provide even more powerful editing features, such as insertion and deletion of entire lines.

5·7   TERMINAL FEATURES, OPTIONS, AND DESIGN

Most modern terminals come with some sort of "setup menu" which allows the operator to select a number of options such as

- the location of typewriter-like tab stops
- full- or half-duplex operation
- communications line baud rate and data format

- whether or not the terminal will electronically simulate "key clicks" and/or "margin bells"
- the number of characters per line
- character or block mode operation

A comprehensive list would be very long, and each terminal manufacturer's models will implement their own peculiar subset. All these options (combined with cursor movement, editing features, and the ability to switch among multiple character sets) makes the modern electronic terminal a very complex device. The pressure of competition among manufacturers has led to an ever increasing list of features. As a result of increasing complexity and the decline in the price of semiconductor devices, almost all modern terminals are internally controlled by microprocessor chips.

What are the basic components of an electronic terminal?

- a keyboard
- a CRT display
- a communications port controlled by a UART
- a microprocessor chip to coordinate the operation of all of these

What is the difference between this and a personal computer? Not much. The primary difference is that in devices called "terminals," the microprocessor's program is stored on read-only memory (ROM) chips. The data contained in a ROM is installed at the factory when the terminal is built. As a result, the program of the terminal's embedded microprocessor cannot be changed by the user of the terminal.[1] For obvious reasons these programs are called *firmware*.

The personal computer, on the other hand, contains a device such as a floppy-disk drive which allows its user to load a variety of programs from external media. These programs are called *software*. The memory that will contain these programs must have the ability to be written (at the time the program is loaded) as well as to be read (at the time the program is executed); therefore, RAM (Random Access Memory) chips must be used. Suppose we now purchase a floppy disk containing software that, when loaded into our personal computer, makes it behave as follows:

- Whenever a key is struck, its ASCII representation is sent, by the software, to the UART transmitter which places it on the communications line.
- Any data pattern received by the UART from the communications line is interpreted by the software as an ASCII character and its equivalent symbol is displayed on the screen.

---

1. As is the case with most classes of electronic equipment produced by multiple vendors, there are exceptions to every general statement. There are some devices called terminals whose programs are modifiable by the user. These, nevertheless, are not personal computers. The user-defined programs usually remain loaded for long periods of time and typically provide communications functions only.

Our personal computer now behaves exactly like a character mode terminal. What our PC is doing is called *terminal emulation*. Block mode terminal emulation software is also available.

As a final comment, there is an enormous variety of CRT terminals currently on the market. Inexpensive ones with few features have been called *dumb terminals* or "glass teletypes." More expensive ones have been called *smart terminals*. However, the variety has increased, and reduced hardware costs have allowed features previously considered smart to be incorporated into dumb devices. This book will therefore make no attempt to define exactly which features differentiate smart from dumb devices.

5·8 ASCII CONTROL FUNCTIONS IN TERMINALS

A number of the ASCII control characters (columns 0 and 1 in Figure 5–3) are related to the operation of terminals. These will be described here.

BEL (CTRL G)

As in Baudot, this character is an audible signal to attract the attention of the terminal operator. The mechanical bell has generally been replaced by an electronic beep from a small speaker—another small example of technological advance and aesthetic decline.

Format Effectors

Format effectors are the nonprinting control characters which determine where the printable characters that follow will appear. On a CRT they move the cursor. On a hard-copy device they move the print mechanism relative to the paper. Programmers sometimes refer to these characters as "whitespace." The ASCII format effectors are

1. *BS (CTRL H): Backspace.* Move left one character position in the current line. In many CRT terminals BS will also erase the character at the new position.
2. *HT (CTRL I): Horizontal Tab.* Move right in the current line to the next character position that has been defined as a tab stop.
3. *LF (CTRL J): Line Feed.* Move down one line position.
4. *VT (CTRL K): Vertical Tab.* Move down in the current page to the next line position that has been defined as a vertical tab stop.
5. *FF (CTRL L): Form Feed.* Move down to the first line position (top-of-form) in the next page. Typical form length (11″ paper) is sixty-six lines.

6. *CR (CTRL M): Carriage Return.* Move left to the first character position of the current line. Many terminals can be configured to automatically produce the sequence CR LF when the carriage return key is pressed.
7. *SP: Space.* Move right one character position in the current line. Since SP is in column 2 of the ASCII chart, it is not formally a control character.

Device Controls: DC1–DC4 (CTRL Q–CTRL T)

The designers of ASCII included four *device control* characters without defining specific functions for them. The intent was to accommodate the needs of a variety of equipment types without adding a large number of characters to the code set. Manufacturers could use DC1–DC4 as they saw fit for their own equipment.

The characters DC1 and DC3, however, have become a de facto standard for an important communications function called *flow control*. Consider the situation (illustrated in Figure 5–7) where we have a computer sending a long data file to a printer via a communications line. At the printer, characters are temporarily stored in a buffer where they wait to be picked up by circuits that drive the print mechanism. The buffer is a small memory that may have roughly 128 bytes of storage capacity (this is approximately the number of characters in a full line of text). If data from the computer is coming in bursts, the presence of the buffer allows a few characters to momentarily stack up while the print mechanism catches up. However, if the computer consistently transmits faster than the printer can print, the buffer will ultimately overflow and data will be lost.

How do we solve this problem? Suppose the specification sheet for our printer claims it can print 30 char/s. If our ASCII line format is 7 data bits, odd parity, and 1-bit stop interval, we know that BITS_PER_CHAR = 10. We could therefore set the line speed at 300 bits/s and be sure that our communications line would never send more than 30 char/s (which our printer supposedly can handle), but we must read specifications carefully. The 30-char/s specification

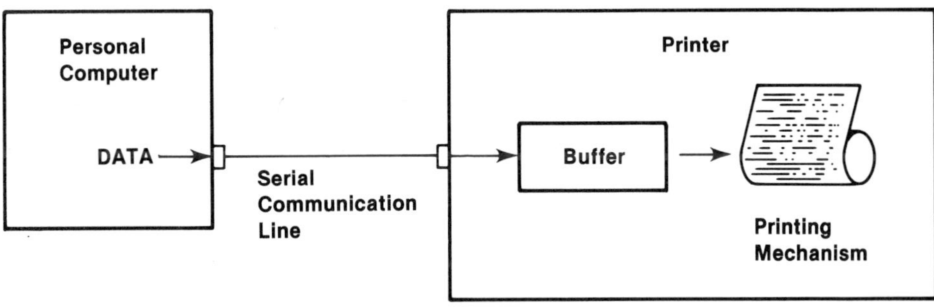

FIGURE 5–7 Data Transfer to a Serial Printer

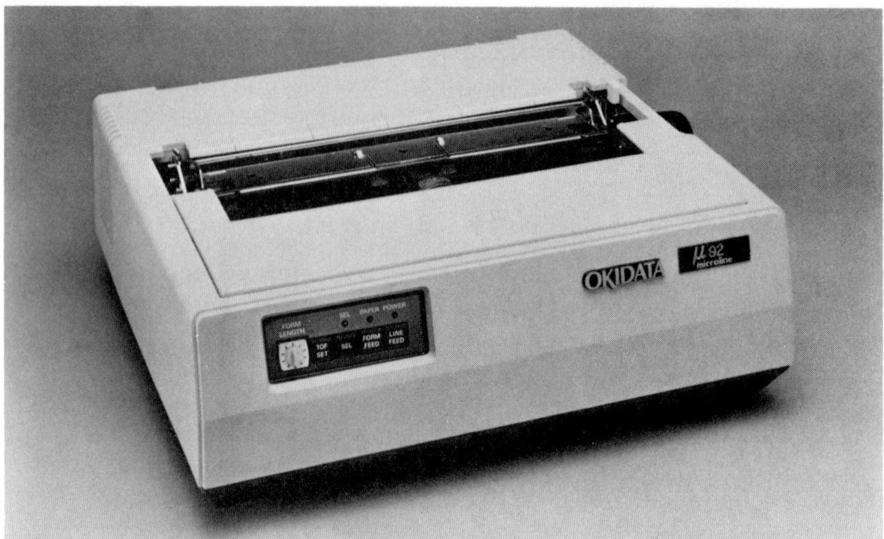

A Typical Printer for a Small Computer System
Courtesy of Okidata

refers to the rate at which text can be printed. We could send the printer a single control character, such as *form feed,* which will require substantially more than 1/30 of a second to execute. Reducing the line speed to 300 baud will not guarantee that the buffer will never overflow.

A more sensible solution is to use a flow control procedure to allow the printer itself to determine the rate of data transfer from the computer. The ASCII characters DC1 and DC3 are so frequently used for this purpose that many manufacturers have renamed them:

$$DC1 = XON \quad \text{(Transmission on)}$$
$$DC3 = XOFF \quad \text{(Transmission off)}$$

The procedure operates as illustrated in Figure 5–8, which is a plot of the instantaneous number of characters in the buffer as a function of time.

- The buffer remains empty until the transmission from the computer starts. We will assume that the baud rate is much greater than the rate at which text can be printed and the number of characters in the buffer will climb rapidly.
- When the buffer is approximately 75% full, the printer will send an XOFF character (DC3) back to the computer. As soon as the computer's program recognizes the XOFF, it will stop transmitting. There may be a slight delay in the computer's response, and this is why the XOFF threshold must be set somewhat lower than 100% of the buffer capacity.

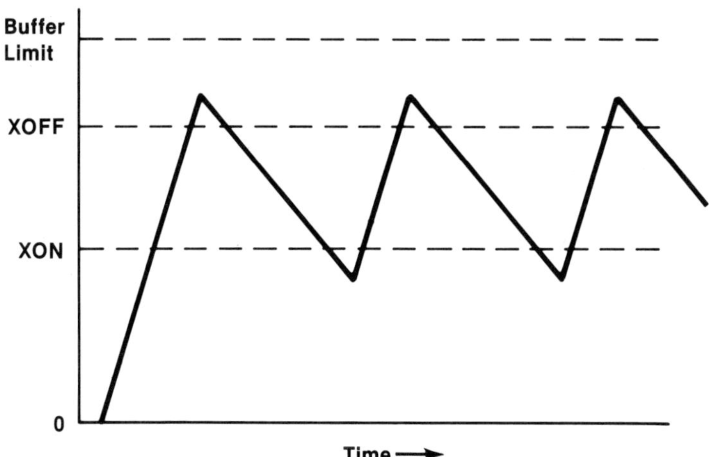

FIGURE 5–8   Buffer Occupancy in a Flow Controlled Printer

- Now that transmission has been temporarily halted, the print mechanism will start to deplete the buffer. The slope of the descending part of the graph is not as steep as the ascent. At some point, buffer occupancy will be down to approximately 50%, and the printer will send XON (DC1) to the computer. The computer will then resume transmission until we hit the XOFF threshold again.

There are many advantages to this procedure. It is self-regulating. The printer operates at its maximum speed regardless of the content of the data file. If we have several different printers that can be attached to our communications port, the baud rate can be set to match the speed of the fastest printer and can remain at this setting regardless of which printer is currently in use. This same procedure is often used by operators of CRT terminals to stop long data transmissions from scrolling off the top of their screens.

There are two assumptions implicit in this discussion of flow control:

- the communications line was full-duplex
- the computer is programmed to recognize and respond to XOFF and XON

Flow control is an important concern in data communications, and we shall see several more examples in the remainder of this book.

Alternate Alphabets: SO/SI (CTRL N/CTRL O)

The *shift-out* (SO) and *shift-in* (SI) control characters behave very much like the figure and letter shifts in Baudot. In terminals that support SO and SI, two different printable character sets or alphabets are available. Normally the hexadecimal values of printable characters are associated with the symbols defined in columns 2–7 of Figure 5–3. If shift-out is received, the graphic interpretations of the hexadecimal values in columns 2–7 are redefined. This alternate symbol set remains in effect until shift-in is received, at which time the normal symbols are reinstated. The alternate symbol set may contain such things as

- characters from non-English alphabets
- mathematical symbols
- intersecting lines and corners of various widths that are useful for creating tables of numbers and bar graphs
- custom symbol sets defined by the user

The SO/SI codes allow the ASCII printable symbol set to be expanded without increasing the number of data bits in the basic character format. Since these special symbols are typically a small percentage of the total text, the "overhead" imposed by the occasional necessity to insert a shift character is not very high.

Inventing New Control Functions: ESC (CTRL [)

It is obviously impossible to anticipate the directions in which technology will advance. At the time ASCII was published, CRT terminals were in a very rudimentary stage of development. As new features were added there was a corresponding need to create new command and control functions. This capability is provided by the *escape* (ESC) character. ESC indicates that the next few characters are to be treated as a multicharacter command sequence rather than ordinary text. The length of the command strings vary. The command may be only the single character following ESC or it may be a string of as many as seven characters. Equipment designers must select unique character strings for each command. *Delimiter* characters must be included to allow the receiving device to determine exactly where the ESC sequence ends. After the end of the ESC sequence, subsequent characters are once again interpreted as normal text.

The keys on a terminal that generate ESC sequences are typically the ones that are not found on the keyboards of conventional typewriters. For example, many character mode terminals have cursor movement keys which are labeled with arrows indicating the direction of motion. Each of these keys may produce a sequence such as

Cursor up:	ESC [A
Cursor down:	ESC [B
Cursor right:	ESC [C
Cursor left:	ESC [D

When the operator hits one of these keys, one of the previous sequences is transmitted to the host computer. If we are operating with remote echo, the same sequence is sent back to the display section of the terminal. Display firmware in the terminal recognizes the sequence and moves the cursor in the appropriate direction. If the application program in the host computer is keeping track of the operator's cursor position, it too must be able to interpret these sequences.

There are also ESC sequences that are unavailable from the keyboard but can be sent from the host computer. For example, suppose the program wants to position the cursor to a specific location on the screen. There may be a sequence such as

$$\text{ESC} \ [\ \langle r \rangle \ ; \ \langle c \rangle \ f$$

This is a 6-character sequence in which $\langle r \rangle$ and $\langle c \rangle$ indicate the row and column where the cursor will be placed. In addition to cursor motion, ESC sequences provide such commands as

- erasing all or part of the line in which the cursor currently resides;
- erasing the entire screen or everything from the current cursor position to the end of the screen;
- editing functions such as inserting blank spaces at the current cursor position or deleting the character at the current cursor position;
- inserting blank lines or deleting existing lines of text;
- indicating to the display that the current cursor position is the beginning (or end) of a protected field;
- indicating to the display that a new video attribute will take effect at the current cursor position;
- allowing the host computer to modify terminal setup parameters such as tab stops, local/remote echo, number of characters per line, and so on.

Each of the terminal's function keys can also generate a unique ESC sequence. These keys are typically used to provide convenience features for the terminal operator, and, for these keys, the host computer will not echo the exact sequence produced by the keystroke. For example, if the host computer were running a language processing program such as BASIC, the various function keys could allow the operator to produce some commonly used commands with a single keystroke. The following table shows the ESC sequences that a hypothetical set of function keys could send to the host computer and the text sequences that the language processor could send in response:

FUNCTION KEY	SEQUENCE SENT TO HOST	SEQUENCE BACK TO TERMINAL
F1	ESC O W	SAVE
F2	ESC O X	LOAD
F3	ESC O Y	LIST
F4	ESC O Z	RUN

The actual list, of course, would be much longer. Other application programs (a word processor, for example) will respond to the same function keys with a totally different set of text sequences.

It is obvious from these paragraphs that the terminal firmware and the host-computer program must agree on the content of the ESC sequences and their meanings. Unfortunately, the ESC sequences for a given function will vary among terminal models and not all functions will be present in every terminal. Standards have been published for ESC sequences but they are not universally followed. Users must carefully check for compatibility in this area before attempting to use a new terminal type with existing programs.

Error Control: ACK/NAK (CTRL F/CTRL U)

We have seen that character parity can be used to detect errors in received messages. If an error is detected, what can the receiver do? It is a common practice to design communications equipment such that long messages are broken up into blocks, and the receiving equipment must respond to each block. The procedure is illustrated in Figure 5–9, where messages and responses are indicated by arrows crossing the "communications channel", and time advances as we move down the diagram. (This is the first of many such diagrams we shall present.)

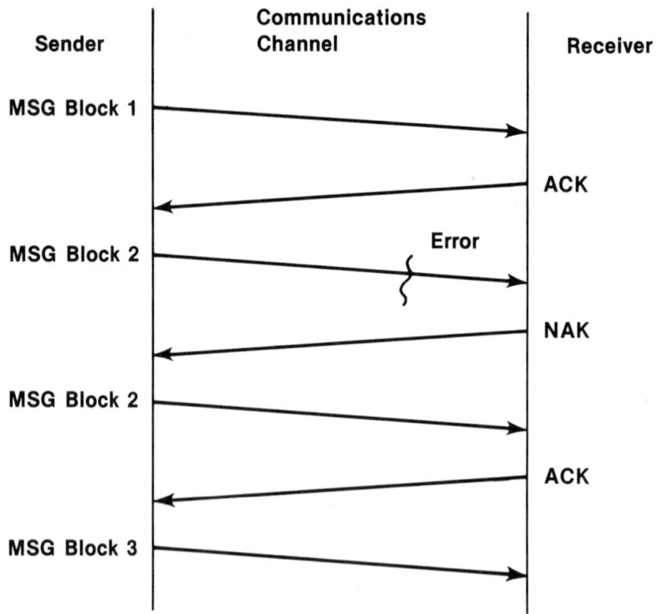

FIGURE 5–9   ACK/NAK Error Control

The sender transmits the first message block and waits for a response. This block was received without errors and the receiver's response is the ASCII control character ACK (acknowledge). After recognizing ACK, the sender is permitted to transmit the second message block. In this case, assume there was a transmission error which was detected by the receiver. The receiver therefore responds with NAK (negative acknowledge). When the sender sees the NAK, it will retransmit message block 2 which is now received without errors. The procedure is repeated until the sender runs out of data. Under this procedure the sender cannot discard a message block from its memory just because it has been transmitted once.

There are many subtle problems associated with error control procedures. For example:

- What does the sender do if there is no response at all?
- How many times does the sender attempt to retransmit if every response is NAK?

We will return to these interesting questions when we discuss the subject of protocols.

REVIEW QUESTIONS

1. List a sequence of binary patterns from the Baudot code that would cause a teletype to print

    GO 4 IT!

2. On an ASCII keyboard, what hex values would be produced by hitting the following keys while CTRL was held down?
    a. X
    b. H
    c. M
    d. V

3. On an ASCII keyboard, how would you produce the following control characters?
    a. EOT
    b. DC3
    c. CR (list two different ways)

4. On a terminal using ASCII code, we have a simple procedure for generating control characters from the alphabetic characters on the keyboard. Could such a procedure be developed for an EBCDIC terminal? If not, explain why.

5. An auto dealer must create a new entry in his inventory data base every time a new car arrives from the manufacturer. Lay out a data entry screen for this function. Your terminal screen has 80 columns and 24 rows. Draw the screen as it would appear before any data was entered, and for each field specify
   a. field length (number of character positions)
   b. protected/unprotected
   c. video attribute

   How do the protection and video attribute features help the operator do the data entry job? What kind of error checks might you incorporate in the program to make sure the operator is not entering nonsense?

6. Obtain the specification sheet for a commercially produced asynchronous terminal and list the escape sequences you would have to send to
   a. clear the screen
   b. position the cursor to row 3, column 10

7. An asynchronous serial printer can print at a rate of 60 char/s. If we connect it to a 600-baud line with 7 data bits, no parity, one-bit stop interval, will it operate reliably without flow control?

chapter six

THE RS-232C INTERFACE AND MODEM CONTROL

In connection with the IBM Asynchronous Communications Adapter (Chapter 4), we briefly mentioned its voltage interface option. We said that these voltage levels complied with the RS-232C specification. This specification, the final version of which was published by the Electronic Industries Association in 1969, describes the interface between digital devices and modems. The full title of the specification is *Interface Between Data Terminal Equipment and Data Communications Equipment Employing Serial Binary Data Interchange.*

The term *Data Terminal Equipment* (DTE) refers to a digital device such as a terminal or a host computer's communications port. *Data Communications Equipment* (DCE) is the device commonly called a modem, which, by a modulation process, converts digital waveforms into signals suitable for transmission on a phone line or other medium. We shall use the abbreviations DTE and DCE throughout the remainder of the book. The terms are easier to remember if you associate the T in DTE with terminals. The title of the specification also includes the words: "Serial Binary Data Interchange" which is, of course, what we have been talking about since Chapter 1.

RS-232C is one of the few standards that is almost universally observed in the data communications industry. This is a great benefit to users because it allows a given computer or terminal to be connected to modems from a variety of different vendors. Nevertheless, compliance with this specification leaves much room for functional variation. Not all terminals are compatible with all modems.

6·1   THE RS-232C "MODEL" OF A COMMUNICATIONS CIRCUIT

The RS-232C specification actually describes the wires in a cable connecting DTE to a DCE (see Figure 6–1). This is the cable that plugs into the familiar

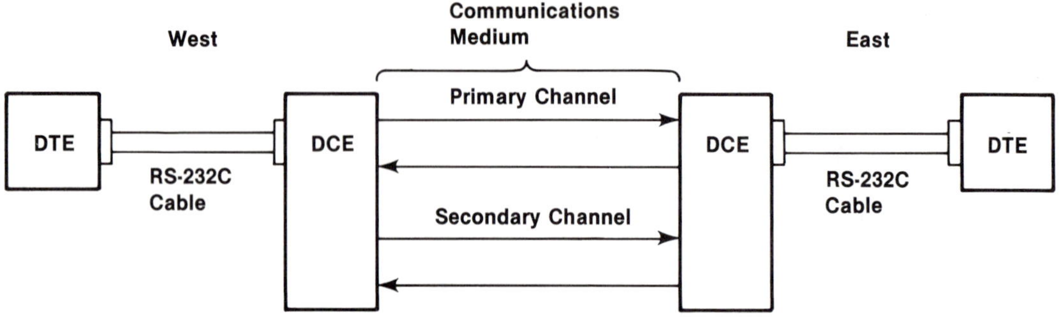

FIGURE 6–1   The RS-232C Model of a Communications Circuit

25-pin "serial port" connector at the back of personal computers and terminals. First, the specification defines the voltage levels and other electrical characteristics of wires in the cable. It then goes on to describe their functions.
These functions fall into four categories:

- data
- grounds
- modem control
- timing

The exact names and a more precise description of the individual wires will
be given later in the chapter. To understand how the modem is controlled by
the DTE, we must understand the "model" or conceptual picture of a communications medium that is used in the RS-232C specification. This model is
shown in Figure 6–1. (Figure 6–1 and many subsequent diagrams in this book
depict two devices connected by a communications medium. For ease of reference we call the devices on the left side of the medium "West," and those on
the right "East.")

The communications medium is composed of a *primary channel* and a *secondary channel*. A channel represents a "pipeline" that has some capacity for
transferring data in either direction. Do not think of these channels as physical
devices or wires. In some cases both directions of all channels may be carried
on a single telephone circuit, and in others the transmission directions of a single channel may be split and transmitted on separate telephone circuits. These
channels are merely concepts required to understand the operation of modem
control wires in the RS-232C cable.

Associated with the channels is the concept of a *carrier*. The carrier is a continuous sine wave signal at a frequency that will easily pass over the communications medium. This carrier is the actual pipeline that allows data to move
from DCE to DCE. The mere presence of the pipeline, however, does not necessarily mean that data is moving through it. In order to send data the carrier

must be modulated. There are several forms of modulation commonly used for data. They are somewhat different than modulation techniques used for analog signals, such as audio, and we will discuss them in detail later.

Each channel actually has two carrier signals, one for each direction of transmission. If both carriers can be active simultaneously, the channel is full-duplex. If only one can be active at any time, the channel is half-duplex.

A channel has a *capacity* which is equal to the maximum number of bits per second it can carry. The secondary channel is not present in all modems, but, if it is, its capacity (speed) is always less than that of the primary. Otherwise there is no conceptual difference between the primary and secondary channels. The RS-232C specification states that it does not apply to communications media with capacities greater than 20 000 bits/s.

RS-232C leaves many choices open to the modem designer. For example:

- either channel may be half- or full-duplex
- the communications medium may be switched or dedicated
- the channels may be synchronous or asynchronous
- the secondary channel may not be present

As a result, the number of RS-232C wires required to control a DCE will vary from model to model. There is probably no single modem that uses all the wires defined in the RS-232C specification.

6·2 ELECTRICAL SPECIFICATIONS

The RS-232C specification describes the electrical characteristics of signals on the cable's wires in great detail. Since we are dealing with binary signals, voltages are restricted to the two regions illustrated in Figure 6–2: *a positive region* (+3 VDC to +15 VDC) and a *negative region* (−3 VDC to −15 VDC). They

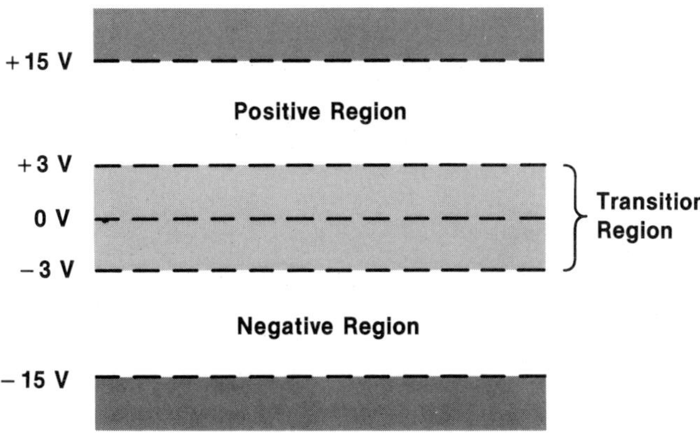

FIGURE 6–2 RS-232C Allowable Signal Voltages

are separated by a *transition region*. Signals are only allowed to enter the transition region during a change of state and may not spend more than 1 ms in this region.

These regions are associated with the binary states of the signal wires in a peculiar way. The interpretation of the voltage level depends on the function of the wire:

	VOLTAGE LEVEL	
WIRE FUNCTION	**Positive**	**Negative**
Data	SPACE (0)	MARK (1)
Modem Control & Timing	On (asserted)	Off (negated)

Notice that, for data signals, MARK is represented by a negative voltage and SPACE by a positive voltage. Oscilloscope observations of asynchronous character waveforms on RS-232C circuits will therefore appear upside down in relation to the illustrations in Chapter 2.

These voltages are obviously not compatible with the input and output specifications of TTL gates. The interface to the RS-232C cable therefore requires special driver and receiver circuits as shown in Figure 6–3. These drivers and receivers are commercially available in standard integrated circuit packages.

RS-232C also limits the capacitance of a signal wire (measured to ground) to a maximum of 2500 pF. For typical wire spacing and insulating materials, 2500 pF is the equivalent of approximately 50 ft of cable. Unless we use special cables with low capacitance insulating materials, 50 ft is the maximum separation between DTE and DCE.

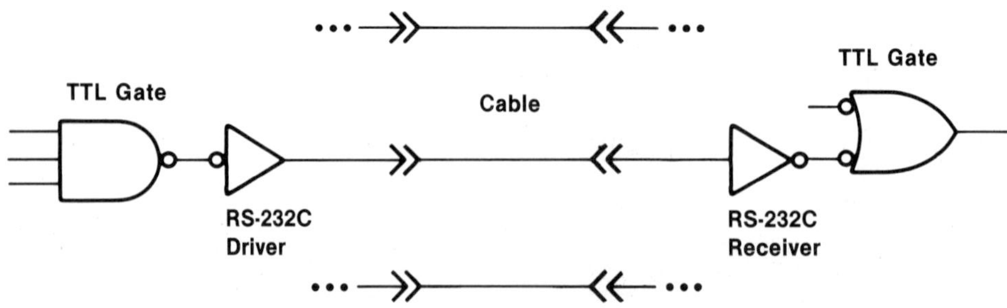

FIGURE 6–3 RS-232C Interface Circuits

6·3 CIRCUIT DESCRIPTIONS AND FUNCTIONS

Figure 6–4 lists the pin numbers in the RS-232C connector and their associated signal names. It also shows signal direction and lists the abbreviations that will be used in this book. (These are not the same as the formal abbreviations listed in the RS-232C document.) Notice that there are two ground wires. Pin 7 (SIGNAL GROUND) is the actual current carrying common return for all data, control, and timing signals. Pin 1 (PROTECTIVE GROUND) is sometimes used to prevent electric shock hazards and reduce circuit noise by establishing electrical contact between the DTE chassis and the DCE chassis.

Data for the primary channel is carried by the TRANSMIT DATA and RECEIVE DATA wires. In RS-232C the terms "transmit" and "receive" are always used from the DTE's point of view. Since there are separate transmit and receive wires, RS-232C is designed to accommodate full-duplex transmission. If, however, the modem or the communications medium is half-duplex, these wires will not be active simultaneously.

Pin #	Signal Name	Direction DTE	DCE	Abbreviation
1	PROTECTIVE (FRAME) GROUND			
2	TRANSMIT DATA	→		XMT
3	RECEIVE DATA	←		RCV
4	REQUEST TO SEND	→		RTS
5	CLEAR TO SEND	←		CTS
6	DATA SET READY	←		DSR
7	SIGNAL GROUND (COMMON RETURN)			GRD
8	CARRIER DETECT	←		CAR_DET
9	—			
10	—			
11	—			
12	SECONDARY CARRIER DETECT	←		SEC_CAR_DET
13	SECONDARY CLEAR TO SEND	←		SEC_CTS
14	SECONDARY TRANSMIT DATA	→		SEC_XMT
15	TRANSMIT CLOCK (DCE SOURCE)	←		XMT_CLK
16	SECONDARY RECEIVE DATA	←		SEC_RCV
17	RECEIVE CLOCK	←		RCV_CLK
18	—			
19	SECONDARY REQUEST TO SEND	→		SEC_RTS
20	DATA TERMINAL READY	→		DTR
21	SIGNAL QUALITY DETECTOR	←		SQD
22	RING INDICATOR	←		RI
23	DATA RATE SELECTOR	→		DR_SEL
24	TRANSMIT CLOCK (DTE SOURCE)	→		XMT_CLK
25	—			

FIGURE 6–4 RS-232C Connector Pin Assignments

In the next few sections we describe the modem control and timing wires of the RS-232C interface. These signals fall into several functionally related groups. A section will be devoted to each group.

6·4 CARRIER CONTROL AND DATA TRANSMISSION (RTS; CTS; CAR DET)

We previously stated that a channel's carrier had to be present before data could be transmitted. The REQUEST TO SEND (RTS) and CLEAR TO SEND (CTS) wires control this function for the primary channel. RTS is a signal from the DTE to the modem that instructs the modem to activate its outgoing carrier. CTS is the modem's response back to the DTE, indicating that the carrier is on and DTE can proceed with data transmission. A related signal, CARRIER DETECT, comes from the modem to the DTE and indicates that the modem has sensed an incoming carrier from the phone line.

Figure 6–5 illustrates how these signals work together. In this and subsequent diagrams, the abbreviation for an RS-232C control signal followed by an arrow pointing upward indicates the time at which the signal turns on. Similarly, an arrow pointing downward will indicate the control signal turning off. The abbreviation for a data wire followed by a double-headed arrow pointing both up and down indicates the time at which the data wire becomes active and data transmission starts. Similarly, a data signal followed by " – " indicates the time at which the wire becomes quiescent. For the remainder of this chapter we will also assume that the communications medium is a telephone line.

The situation illustrated in Figure 6–5 is a half-duplex circuit. In such cases there must be some prearranged rule to determine which party is allowed to transmit first. Assume that West has this privilege. To indicate its desire to begin data transmission, the West DTE will turn on REQUEST TO SEND. The modem (West DCE) will sense the change in RTS and activate its outgoing carrier signal. West DTE cannot begin to send data until it receives CLEAR TO SEND from West DCE. West DCE does, in fact, turn on CTS after a delay of approximately 150 ms. The reason for this delay will be explained shortly. In the meantime, West DCE's outgoing carrier has arrived at East DCE, which will then turn on CARRIER DETECT. The CARRIER DETECT signal is sensed by East DTE, which interprets this as an indication of the imminent arrival of data from West. The West to East carrier is now active, but there is no modulation yet.

After the 150-ms delay, West DTE senses that its CLEAR TO SEND signal is on. West DTE can now begin to put information on its TRANSMIT DATA wire. This causes West DCE to begin modulating its carrier. The resulting signal is demodulated by East DCE, and the activity on East DTE's RECEIVE DATA wire now reflects exactly what is being sent by West DTE. After some time, West DTE has

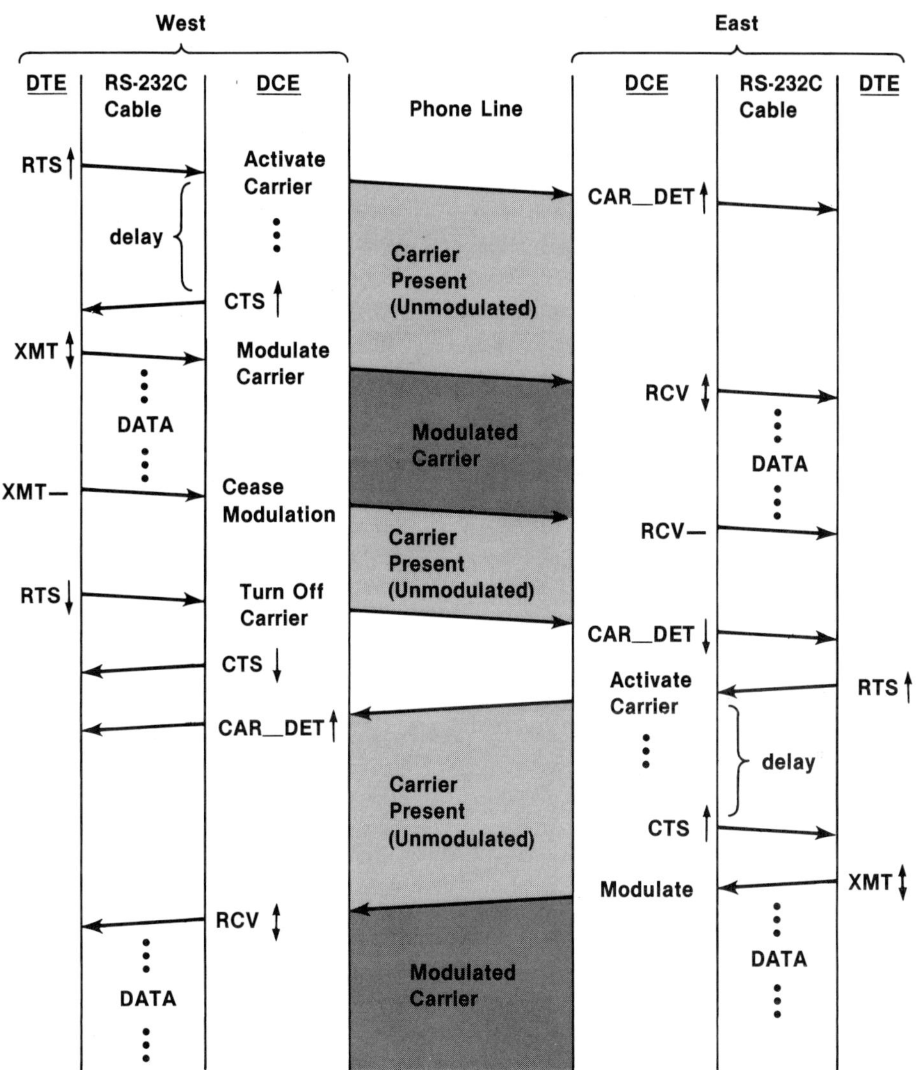

FIGURE 6–5 Half-Duplex Carrier Control

transmitted everything it currently wishes to send, and its TRANSMIT DATA wire becomes quiescent (hold at steady MARK). East DTE's RECEIVE DATA line similarly becomes quiescent. West DTE turns off RTS, and West DCE responds by turning off CTS and removing its carrier from the phone line. East DCE notices the disappearance of an incoming carrier and turns off CARRIER DETECT.

All is quiet once again, and it is now East's turn to transmit. The entire procedure is now repeated in the opposite direction. It is initiated when East DTE turns on its REQUEST TO SEND. This reversal of carrier direction is called a *line turnaround,* and it is a characteristic of all half-duplex communications.

Notice that the instant when East DCE turns on its carrier (in response to RTS from East DTE) occurs almost immediately after the carrier from West DCE has been turned off. As we shall see in the next chapter, on long-distance data calls, echoes of transmitted signals are frequently present. (These echoes are usually not noticeable during voice conversations.) For this reason, at the instant that East DCE activates its carrier in Figure 6–5, West DCE may still be hearing echoes of its own carrier. Since half-duplex modems typically use the same carrier frequency for both directions, the echoes of West DCE's old carrier will interfere with the new carrier from East DCE. We must therefore prevent East DTE from transmitting data until these echoes die off—hence, the delay from RTS to CTS.

The delay associated with line turnaround really slows down half-duplex communications. When operating at high data rates, half-duplex circuits may spend as much time turning the line around as they do transmitting data. For example, consider a 9600-bit/s line transmitting 8-bit synchronous characters. From the equations in Section 2·5 we have the following (since there is no start bit or stop interval for synchronous transmission, BITS_PER_CHAR = 8):

$$\text{CHAR\_PER\_SEC} = \frac{\text{B\_P\_S}}{\text{BITS\_PER\_CHAR}} = \frac{9600}{8} = 1200$$

We can now compute the equivalent number of characters that could have been transmitted during a 150-ms turnaround delay:

$$1200 \text{ char/s} * 0.150 \text{ s} = 180 \text{ char}$$

For reasons that will be explained in a later chapter, synchronous devices generally transmit long messages as a series of short blocks, and transmission rules frequently specify that the receiver must respond to each block. This requires two line turnaround delays for each transmitted block on a half-duplex circuit. Since 180 characters is within the typical range of block sizes used on synchronous circuits, it is easy to see the impact of turnaround delay on overall data transmission speed. (Review Question 6–3 is a more detailed example of this effect.)

The situation for full-duplex circuits (illustrated in Figure 6–6) is considerably simpler because both carriers can be present simultaneously. The question of which end is first to turn on REQUEST TO SEND is irrelevant. In Figure 6–6 we have chosen West to be first. The echo problem does not exist for full-duplex circuits. (Either the East and West carriers are different frequencies, or the telephone circuit may be set up in a manner that eliminates echoes. This will be explained in the next chapter.) Since we do not have to worry about echoes, the delay from RTS to CTS can be minimized. Many modems therefore give the user the option of selecting RTS–CTS delay from a number of possible values. These values usually range from a few milliseconds to approximately 200 ms.

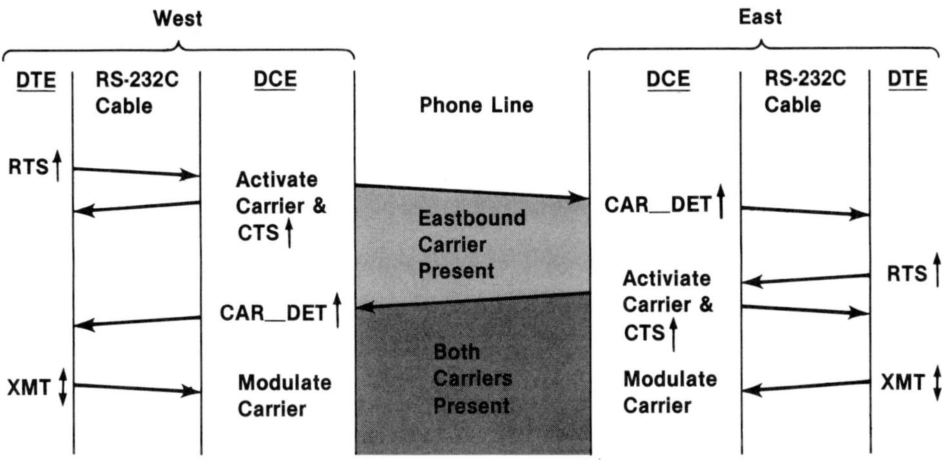

FIGURE 6–6   Full-Duplex Carrier Control

Once the two carriers are established, either end may transmit at any time. If the two DTEs are designed to alternate their transmission activities, we have half-duplex communications over a full-duplex medium. This may seem wasteful, but it is actually done quite frequently. Since both carriers are on constantly there is no need for line turnaround delays, and this alone can make a substantial improvement in the throughput of the circuit.

6·5   THE SECONDARY CHANNEL

In Figure 6–4 there are five wires associated with the secondary channel:

- SECONDARY TRANSMIT DATA
- SECONDARY RECEIVE DATA
- SECONDARY REQUEST TO SEND
- SECONDARY CLEAR TO SEND
- SECONDARY CARRIER DETECT

These all provide, for the secondary channel, functions analogous to those of their similarly named counterparts for the primary channel. By definition, the maximum possible data rate on the secondary channel must be less than that of the primary channel. Many modems do not have a secondary channel at all, and in those that do, it is usually implemented in a special configuration called a *reverse channel*.

Reverse channel modems can switch between the two different states depicted in Figure 6–7. The primary channel (represented by the heavy arrow) is half-duplex and controlled exactly as described in the previous section. The

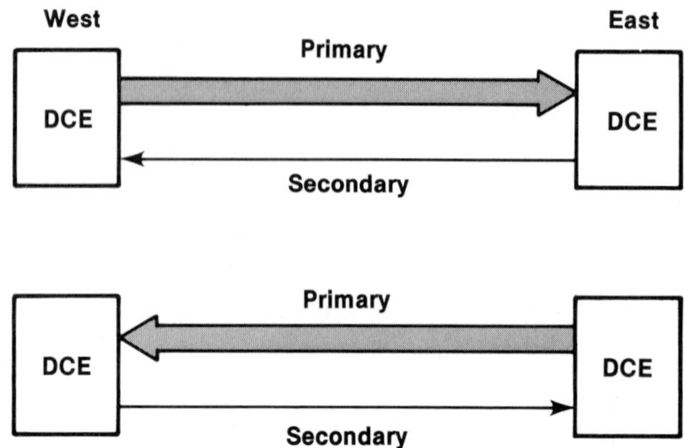

FIGURE 6–7   Alternate States of Reverse Channel Modems

transmitter for the secondary carrier is interlocked with the primary channel in such a manner that, when East's primary carrier is active, East's secondary carrier cannot be activated (even if East's SECONDARY REQUEST TO SEND is on). The same, of course, is true for the West DCE. The result is a sort of "three-quarter–duplex" arrangement where we can have two-way simultaneous transmission, but one direction always has a much higher capacity than the other. One common modem model has a primary channel with a capacity of 1200 baud with a 5-baud reverse channel.

What is the purpose of such arrangements? We have already discussed a few applications where the data rates were highly unbalanced according to direction:

FLOW CONTROL   In the previous chapter we discussed an example of XOFF/XON flow control involving a computer and a printer. We stated that this procedure required a full-duplex channel. Actually the heavy data flow was in the direction of the printer with only an occasional DC3/DC1 control character going in the opposite direction. This application is ideally suited for reverse channel modems. In fact, we do not even need to send control characters in this situation. When the printer buffer is filling, it can turn off its SECONDARY REQUEST TO SEND wire. The computer can interpret the consequent negation of its SECONDARY CARRIER DETECT as a request to suspend transmission.

ERROR CONTROL   The ACK/NAK error control example in the previous chapter is another example of an application with unbalanced data flow. This too is a good candidate for a reverse channel modem pair. In this case we must actually transmit data (ACK or NAK characters) on the secondary channel. After a number of data message blocks have been transmitted by West DCE (with ACK/NAK coming from East DCE), the primary channel can be turned around, allowing East DCE to transmit message blocks while West DCE sends ACK/NAK.

CIRCUIT ASSURANCE   When long data files are being sent on a half-duplex channel, there may be no way that the transmitter can be certain that the data is actually reaching the intended destination. A switched connection may be inadvertently broken, or a hardware failure may occur in the remote modem or DTE. In this situation the transmitting equipment must alert an operator as soon as possible so the problem may be repaired and transmission restarted. If reverse channel modems are available, the presence of the secondary carrier (SECONDARY CARRIER DETECT on) can assure the transmitting device that the connection is intact and the equipment at the remote end is operational.

6·6   AUTOMATICALLY ANSWERING INCOMING CALLS (DSR; DTR; RI)

The next group of RS-232C wires we will discuss is required for the use of modems on switched telephone lines. Modems for switched lines are designed with the ability to answer incoming telephone calls and disconnect them. The DTE can control this procedure in a manner that is totally analogous to the way people use a telephone. A DTE can also control the origination of a call. This will be discussed in the next chapter. The signals involved in answering a call are

DATA TERMINAL READY
DATA TERMINAL READY is a signal from the DTE that is turned on to indicate the DTE's willingness to accept incoming calls. It is turned off to initiate disconnection of the call.

DATA SET READY
DATA SET is yet another name for modem (we now have three: data set, DCE, and modem). When a modem is connected to a switched circuit, DATA SET READY is turned on by the modem to indicate that a call connection has been established. It is subsequently turned off to indicate that the disconnection process is complete.

RING INDICATOR
RING INDICATOR is a signal from the modem indicating that a *ringing signal* is being received from the phone line. It is the means by which the DCE tells the DTE that someone is attempting to call ("the phone is ringing").

DATA SET READY and DATA TERMINAL READY are sometimes present on devices designed for dedicated circuits. In this case, they merely indicate that the DCE or DTE has power applied and is in operational condition. Their use on a switched circuit is illustrated (in Figure 6–8) from the point of view of the West DTE, which is the "called" party. The "calling" party is not shown in Figure 6–8.

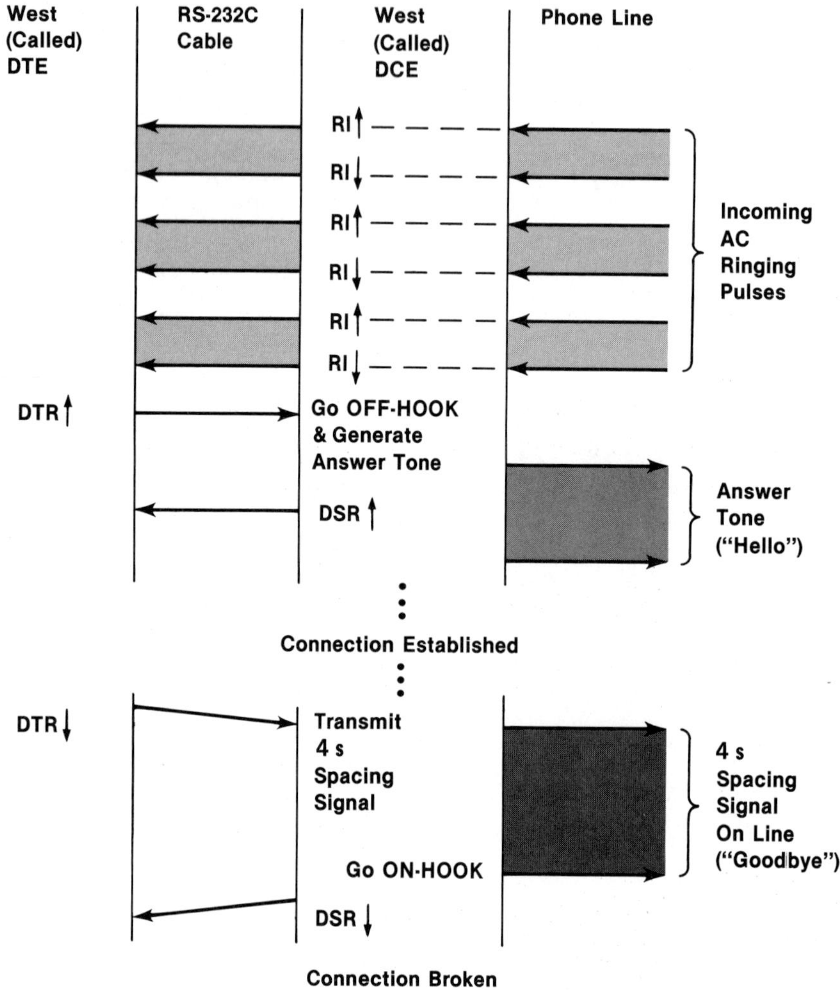

FIGURE 6–8   DTE Controlled Answer and Disconnection of a Phone Call

The ringing signal on a telephone line is a series of pulses of alternating current. The amplitude of the pulses is approximately 100 VAC at a frequency of 20 Hz. The duration of the AC pulse is one second with a three-second interval of silence between pulses. This is the familiar cadence of a ringing telephone. When the ringing signal arrives at West DCE, it turns RING INDICATOR on and off in synchronism with the incoming AC pulses.

West DTE senses the state changes on RING INDICATOR, and, if it wishes to accept the call, turns on DATA TERMINAL READY. When the DCE senses DATA TERMINAL READY, it goes *off-hook*. That is, the modem electrically simulates the event that takes place on a phone line when a person lifts the handset (receiver) of a ringing phone. In the next chapter we will describe the exact nature of a phone going off-hook. In any case the call has now been "answered," and the ringing signals cease. The modem will now transmit an *answer tone*. This is a sine wave (with an approximate frequency of 2000 Hz) that lasts for a few seconds. The purpose of the answer tone is to inform the calling device that the call has been answered. It is analogous to a person saying "hello" after picking up a ringing phone. The answer tone is not necessarily the carrier. It may be a different frequency than the carrier, and, on a half-duplex circuit, REQUEST TO SEND will probably not be turned on at the time an incoming call is answered. After the answer tone has been transmitted, the modem will turn on DATA SET READY to indicate to the DTE that the connection is now established.

Once all this has taken place, carriers can be turned on and data transferred between the called and calling DTEs exactly as we described in the previous sections. At some point one party or the other will want to disconnect the call. In Figure 6–8 the disconnection is initiated by West DTE, but the identical procedure could have been initiated by the DTE at the other end. West DTE will indicate its desire to disconnect by turning off DATA TERMINAL READY. As a result, the modem will transmit a steady SPACE signal for approximately four seconds. The purpose of this signal is to tell the remote DCE: "This call is about to be disconnected, please hang up your telephone." This is analogous to a person saying "good-bye" at the end of a call. East DCE must be designed to go *on-hook* (hang up) when it detects a long SPACE. (This feature is frequently an option on modems and may be disabled by the user.) In any case, West DCE will go on-hook after transmission of the 4-s SPACE is complete. It will also turn off DATA SET READY to inform West DTE that the disconnection process is complete.

The process described is only typical, and in actual modems many variations are possible. For example, the called DTE could continuously maintain DATA TERMINAL READY in the ON state during the interval between incoming calls. In this case the DCE would have answered on the first ring pulse. Also, the long SPACE indicating imminent disconnection could be sent by the DTE rather than the modem or possibly not sent at all. If the long SPACE is not sent, the remote modem must be configurable to recognize some other event as an indication that it must go on-hook (loss of carrier is sometimes used). Otherwise, one end of the call may remain indefinitely in the off-hook condition after the other end has hung up.

6·7   TIMING FOR SYNCHRONOUS CIRCUITS (XMT CLK; RCV CLK)

In Chapter 2 we briefly referred to the synchronous data format. The significant feature of this format was the elimination of idle time between characters. Stated another way, the first bit of a character immediately follows the last bit of the preceding character, and there are no start bits or stop intervals. While this obviously increases the efficiency of line usage, it also presents some technical problems.

A small section of a synchronous transmission waveform is illustrated in Figure 6–9. What we have shown is the distress message "SOS" transmitted in 7-bit ASCII (no parity) and synchronous format. To visualize how this would appear on an oscilloscope, cover the interpretations shown above the waveform and imagine that the tick marks at the bit boundaries are not present. Now imagine this expanded to a message that is 200 characters long. How are we to make sense of this apparently random signal?

In the asynchronous format, the start bit had two functions:

1. It separated characters into recognizable units.
2. It allowed us to synchronize our Receive Shift Register clock to the mid-points of the incoming data bits.

In order to make the synchronous format useful, we must deal with both of these problems. The first falls within the subject of protocols, which we will discuss in a later chapter. The second is handled by synchronous modems and the RS-232C interface.

The synchronous DTE must have a device that performs a serial-to-parallel conversion function just as the UART did for the asynchronous DTE. We therefore need a Receive Shift Register and a receive shift clock. In Figure 6–4 there is a signal from DCE to DTE called RECEIVE CLOCK. This wire is present in the RS-232C interface from all synchronous modems. RECEIVE CLOCK goes from ON to OFF precisely at the midpoint of each bit as it appears on the RECEIVE DATA wire. This is exactly what is required to drive the receive shift register.

We have now handed the problem over to the modem. How does the modem derive the clock? Merely knowing the baud rate is not sufficient, because a small difference in frequency between the transmitting and receiving modems will cause the clock transitions to drift far from the mid-bit points over the

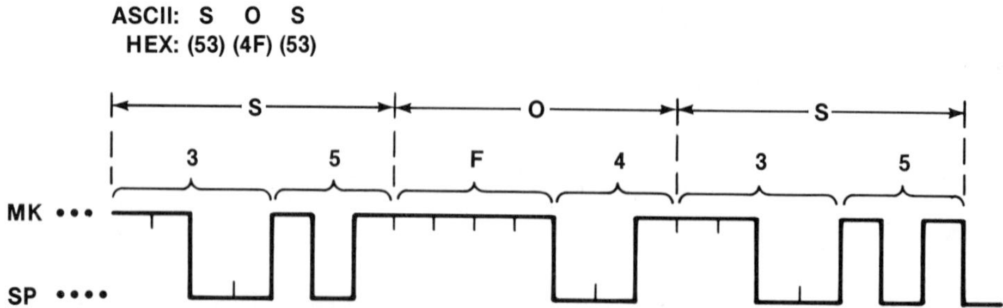

FIGURE 6–9   ASCII Text in Synchronous Format

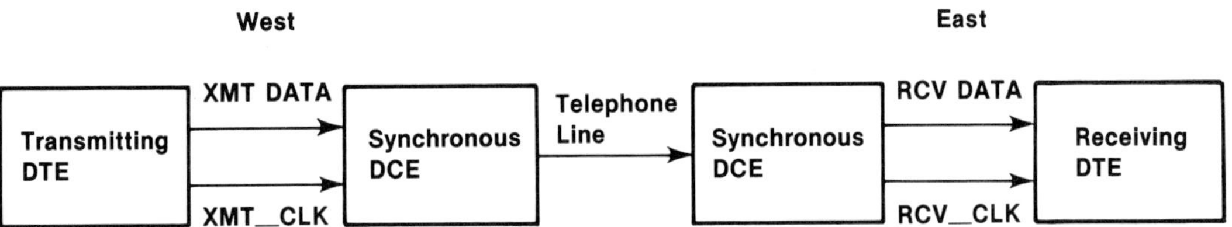

FIGURE 6–10 **Block Diagram of Synchronous Transmission**

duration of a long message. The only true source of bit timing in the synchronous format is the transmitting DTE itself. In Figure 6–4 there is a signal on pin 24 called TRANSMIT CLOCK (DTE SOURCE). This signal is a required input to synchronous modems.

Figure 6–10 is a block diagram for synchronous data being sent from West to East. West DTE provides TRANSMIT DATA and TRANSMIT CLOCK to West DCE. The TRANSMIT CLOCK ticks in exact synchronism with the appearance of bits on TRANSMIT DATA. TRANSMIT CLOCK is therefore the exact signal required to drive West DTE's Transmit Shift Register. We now have precisely timed data entering West DCE and emerging from East DCE. How does the clock signal get across the phone line? Remember that we have only one phone line and only one carrier that is capable of moving information at the full capacity of the primary channel. This leads to a conclusion that is the key to synchronous data communication:

Modulation techniques used for synchronous communications must have the ability to place both data and clock signals on the carrier at the transmit end of the channel and then extract these signals from the carrier at the receive end.

An example of a modulation technique that can do this will be presented in a later chapter.

Actual modem interfaces usually offer an optional variation to the scheme just described. In Figure 6–4 there is a wire on pin 15 called TRANSMIT CLOCK (DCE SOURCE). As we stated previously, the TRANSMIT CLOCK on the RS-232C interface between West DTE and West DCE (Figure 6–10) must be the exact signal that is used to drive the Transmit Shift Register inside West DTE. As long as this is true it really does not matter whether the source of the clock is the DTE or the DCE. Users of synchronous modems are often given the choice of using either TRANSMIT CLOCK wire. A typical arrangement is illustrated in Figure 6–11. With the switches SW1 (DTE) and SW2 (DCE) set to the positions shown, the DTE is the source of the TRANSMIT CLOCK for both the Shift Register in the DTE and the modulator in the DCE. If both switches were flipped to their alternate positions, the DCE would become the clock source. If the switches were in opposite states, of course, we would have a problem.

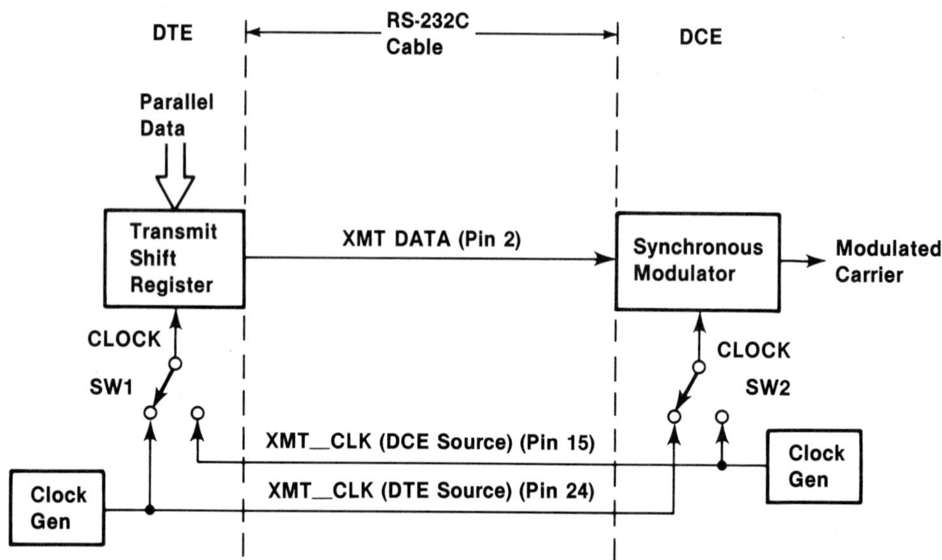

FIGURE 6–11   Synchronous Modem TRANSMIT CLOCK Options

There is a small advantage to using the DCE as the source of TRANSMIT CLOCK. We could then switch to a modem with a different baud rate, and the DTE would not have to be reconfigured.

REVIEW QUESTIONS

1. Would you expect to find the RECEIVE CLOCK wire on the RS-232C connector to an asynchronous modem? If not, why?

2. Which RS-232C circuits would be absolutely necessary for a DTE to control a half-duplex, synchronous modem (no secondary channel) operating on a dedicated circuit?

3. West host computer is transferring a large data file to East host computer on a half-duplex, 4800-bit/s synchronous line. West host computer sends 8-bit characters in blocks of 128 characters. After each block the line is turned around and East host computer sends a 3-character acknowledge message. The line is then turned around again and West host computer sends the next block, and so on.
   a. If the line turnaround time is 150 ms, what percentage of the total time is spent on turnarounds?

b. Over a long period of time what is West host-computer's average transmission rate in characters per second?

c. How much would West host computer's average transmission rate improve if a 100-bit/s reverse channel was available? (Assume that West host computer must still wait for the acknowledge message before it can send the next block.)

d. How much additional improvement in average transmission rate would we realize if a full-duplex channel was available?

4. Draw the waveform of the following EBCDIC sequence in synchronous format (control characters are in "( )"):

    (SYN) (SYN) (SOH) A

What do you think is the function of the SYN control character? Would 66 (hex) be a suitable pattern for a SYN character?

5. What is the function of the RS-232C TRANSMIT CLOCK signal?

6. Redraw the disconnect sequence at the bottom of Figure 6–8 as you imagine it might appear if the long-SPACE disconnect had been originated by East instead of West.

7. When DTEs are close enough to be connected without using modems, RS-232C cables with wiring similar to that shown are frequently used. Label the pins on the both sides of the cable with their RS-232C names and explain the significance of these connections.

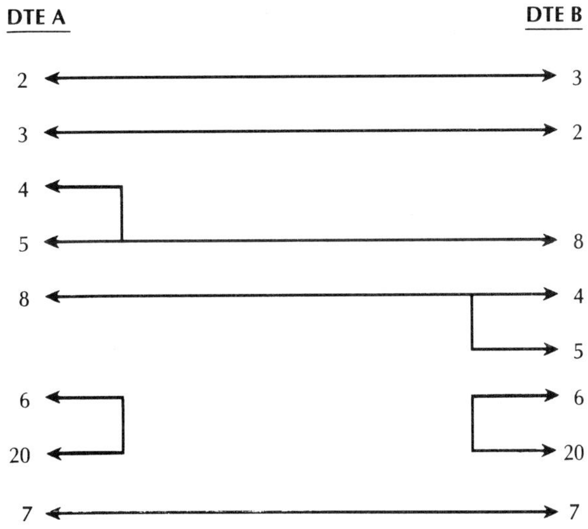

chapter seven

THE TELEPHONE SYSTEM AND DATA CALLS ON SWITCHED CIRCUITS

Several chapters ago we started exploring data communications by studying the operation of the UART chip. We have since added successive layers onto this foundation: the interface PCB; terminal hardware; the RS-232C interface; and modems. We have also seen how digital devices can exchange data and answer incoming phone calls. Before we can proceed to subjects such as automatically dialed calls and the analog side of modems, we must gain a superficial understanding of the telephone network. The intent is not to make the student an expert on the telephone system, but to clarify the design constraints imposed upon communications equipment by a medium that was intended for voice transmission.

7·1   THE TELEPHONE

Figure 7–1 is a simplified schematic of a conventional telephone. The connection between your telephone and the worldwide interconnecting network is a *two-wire circuit* (signal and return) that the phone company calls a *subscriber loop*. All signals that are necessary for telephone operation must pass across the subscriber loop. The variety of signals your telephone must send and receive is surprisingly large:

- ringing pulses
- off/on-hook indications
- dial tone
- dial digit signals (rotary or tone)
- busy signals
- voice (in two directions)

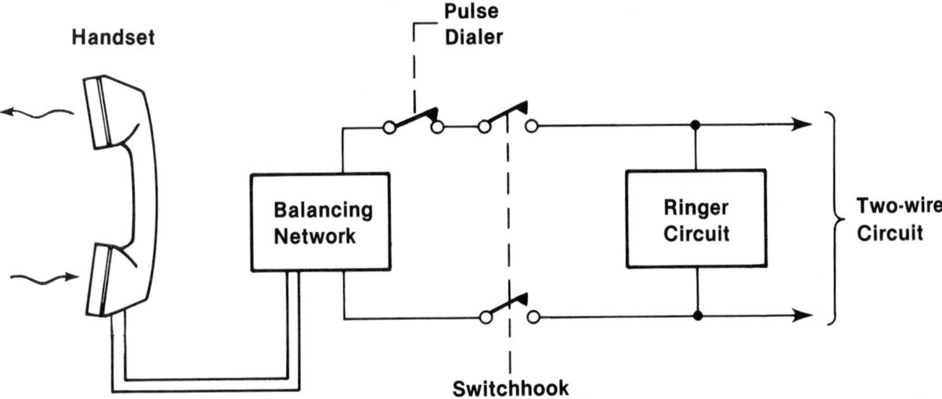

FIGURE 7–1 A Simplified Schematic of a Telephone

- signals indicating that the called party's phone is ringing
- miscellaneous other signals, such as the loud wail telling you that you forgot to hang up an extension phone

The first device we encounter as we proceed from the subscriber loop into the telephone (see Figure 7–1) is the *ringer*. This is the device that responds to the AC ringing pulses described in the previous chapter and produces an audible ring. Immediately to the left of the ringer, and in parallel, is the *switchhook*. The switchhook has two contacts which are in the open position when the *handset* is in the cradle of the telephone (that is, when the phone is hung up). In this condition the telephone is said to be *on-hook*. This term originated from early telephones which, like modern wall phones, had a hook-like device from which the handset was suspended when the phone was not in use. When someone lifts the handset the telephone is said to be *off-hook*.

The nearest telephone company facility to a subscriber's home or place of business is called an *End Office*. The far end of the subscriber loop enters the End Office, and, at this point, the loop may be connected to a variety of devices that allow access to the rest of the network. One of these devices is a direct-current voltage source (called the *line battery*) connected to all subscriber loops. There is also a current sensor in each subscriber loop.

When the phone is on-hook, the only device in the telephone that is connected across the subscriber loop is the ringer. The ringer is designed to respond to the AC ringing pulses, but it will not pass any current in response to the DC line battery voltage. When the handset is lifted (goes off-hook), the switchhook contacts close, and the *balancing network* appears in parallel with the ringer (the *pulse dialer* contact is normally closed at this time). The balancing network has a DC resistance. The line battery will therefore produce a direct current in the line which is sensed at the End Office as an indication of the off-hook condition. In response to the subscriber going off-hook, a device in the End Office which is capable of storing and interpreting dialed digits is attached to the subscriber loop. At the time of attachment the subscriber begins to hear *dial tone*. At the End Office there are a limited number of these digit collection devices (far fewer than the total number of subscribers). If you go off-hook at a time when the End Office is busy, the dial tone may be delayed until one of these devices becomes available.

The next event is dialing by the subscriber. The pulse dialer contact shown in Figure 7–1 is a feature of the old-fashioned rotary dial phone. It works as follows:

- The dial has ten holes, each of which represents a possible dial digit. The subscriber puts his finger in one of these holes and rotates the dial clockwise until his finger hits a stop. The amount of rotation allowed by this operation is a function of the digit selected by the subscriber. The pulse dialer contact has remained closed during this operation.
- The subscriber removes his finger from the dial and allows it to freely rotate counterclockwise to the home position. During the counterclockwise rotation (which takes place at a mechanically controlled rate) the dial operates a cam mechanism that repeatedly opens and closes the pulse dialer contact, briefly interrupting the DC loop current. The number of interruptions that take place during the counterclockwise rotation is equal to the digit selected by the subscriber (except for the digit 0, which produces ten interruptions). These interruptions are separated by approximately 100 ms, and the duration of each interruption is approximately 60 ms.

Figure 7–2 is an illustration of a subscriber-loop current waveform that summarizes the previous discussion.

More modern telephones use tone (or push button) dialing. The formal name for this concept is *Dual Tone Multifrequency* (DTMF) dialing. In this case the pulse dialer contact is replaced with an oscillator circuit that produces a unique combination of two audio frequencies (hence, "dual tone") each time

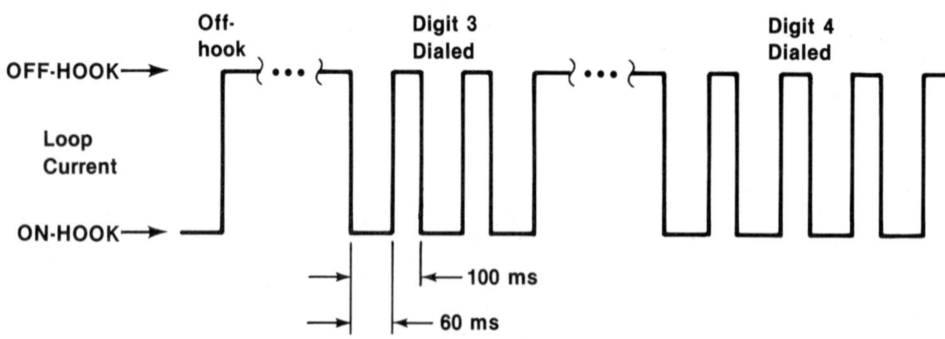

FIGURE 7–2   Subscriber Loop Current at the Beginning of a Call

a button is depressed. The frequencies are selected from the matrix illustrated in Figure 7–3. For example, when 5 is pressed, a row frequency (1336 Hz) and a column frequency (770 Hz) are simultaneously generated and transmitted on the subscriber loop. Note that the signal which appears on the subscriber loop is the sum of the amplitudes of the two tones. No summing of frequencies takes place. Pressing the digit 5 does not produce any measurable energy at a frequency of 2106 Hz (1336 Hz + 770 Hz).

The DTMF button array includes two extra symbols (∗ and #) which are never used for dialing. Since DTMF tones are readily transmitted to a remote listener (or listening device) while a call is in progress, it is possible to use a DTMF telephone as a simple computer terminal. The symbols ∗ and # can be used to separate or terminate numeric data entries for such an application.

The most obvious advantage of DTMF over pulse dialing is speed. A DTMF button's output will be reliably interpreted by End Office equipment if it is held down for only 100 ms. The time between digit entries is also reduced. This is an advantage for both the dialer and the phone company. As we stated previously, the End Office equipment that accumulates dial digits must be shared by a large number of subscribers. The shorter the time interval that one of these devices will be kept busy by a single call, the fewer the number of devices that will suffice for a given number of subscribers.

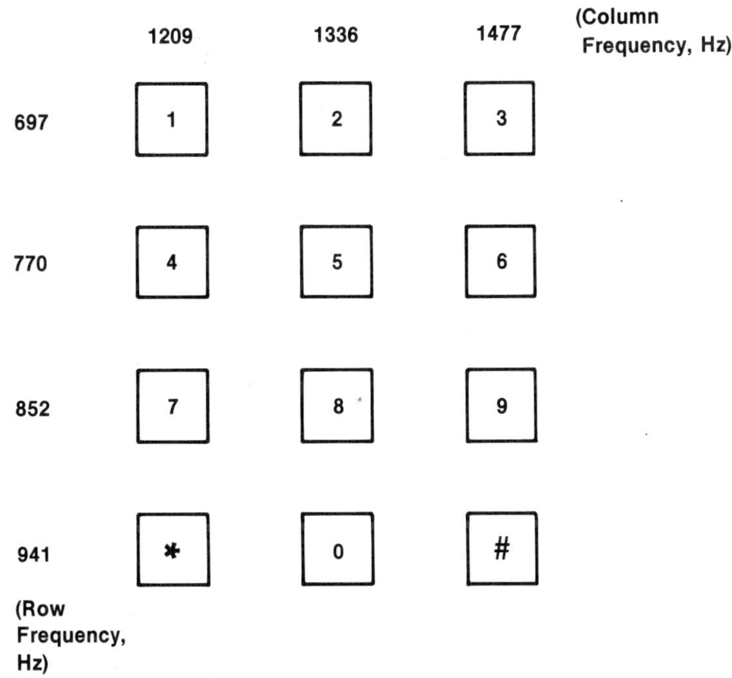

FIGURE 7–3   DTMF Dialing Frequencies

To get an idea of how important dialing speed is to telephone operating companies, consider the following. At the time that the area code system was developed for long-distance calls, most phones in the United States had rotary dials. The 3-digit area codes were selected according to strict rules (for example, the middle digit has to be either 0 or 1). Only a limited number of codes were therefore available and some were obviously faster to dial than others. Look at the area code map in your telephone directory and think about the relationship between the volume of calls to a particular location and the time required to dial its area code on a rotary phone. For example:

> New York, NY: area code 212
> Amarillo, TX: area code 806

The other components of the telephone—the *balancing network* and the *handset*—are concerned with the bidirectional transmission of voice signals. As we previously stated, the telephone network limits audio signals to a bandwidth of 300 Hz to 3300 Hz. This bandwidth limitation is the single factor that has the strongest impact on the design of telephone modems.

The termination of a call is detected at the End Office by the drop in loop current when the subscriber goes on-hook. The on-hook condition is usually required to last for a second or two before the connection is actually broken. We do not want End Office devices that monitor for on-hook to be activated by momentary glitches or the brief interruptions caused by pulse dialing. This is useful for data communications because situations arise in which we want to switch the subscriber loop from one device to another without disconnecting the call. We will see examples of this later in the chapter.

7·2   SWITCHED CALLS

In Figure 7–4 we have illustrated a telephone switching system for three, very small, neighboring towns. Each town has a few *subscriber telephones* and an end office (EO A, EO B, EO C) which is the terminus for subscriber loops. The lines connecting the End Offices to each other are called *trunks*. There may be many trunks between a pair of offices. Each End Office is assigned a 3-digit *prefix,* and each phone has a 4-digit number. These combine to form the familiar 7-digit telephone number or *network address* of a subscriber. This closely resembles the chip-select/register-address scheme we discussed in connection with UARTs. Notice that the appearance of a subscriber numbered 1234 on both EO A and EO B does not cause any problems because they are separated by their End-Office prefixes. All subscribers on a single End Office must obviously have unique numbers.

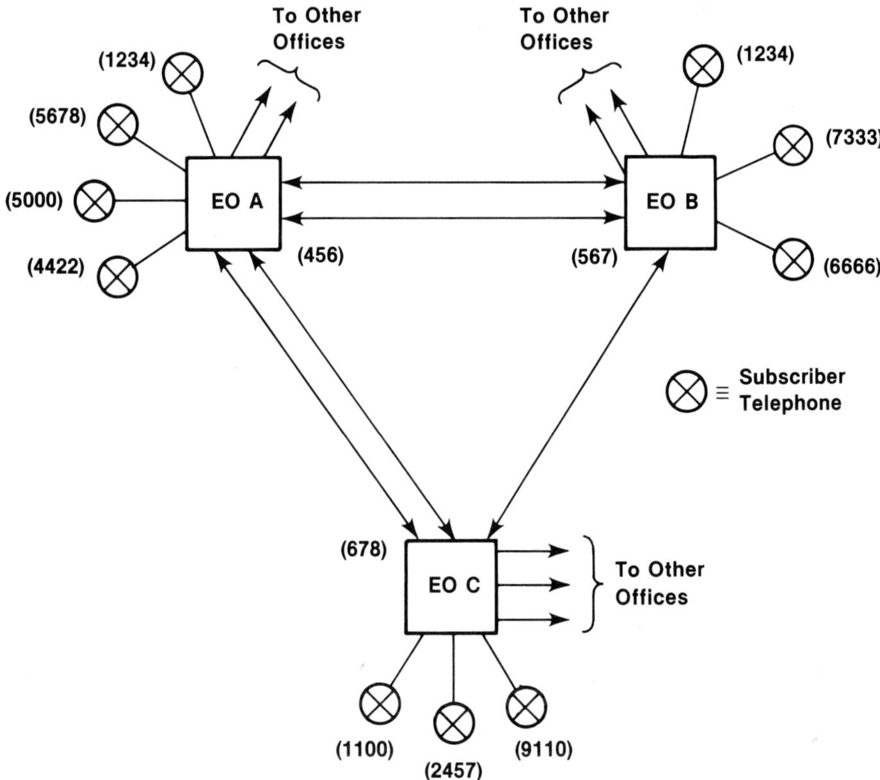

FIGURE 7–4 Local Telephone Switching

Suppose subscriber 5000 on EO A wants to call subscriber 4422 on the same End Office. He goes off-hook and dials 456-4422. The End Office recognizes its own prefix and attempts to set up a connection between the two phones. If, at the moment, 4422 happens to be talking to someone else, the caller will get a busy signal. The busy signal is a combination of two audio tones (480 Hz and 620 Hz) which pulse at a one-per-second rate. If 4422 is not busy EO A will

- establish an electrical connection between the two subscriber loops;
- send ringing signals to the called party (4422);
- send a signal that sounds like ringing back to the calling party (5000).

The sound the calling party hears (*ring back*) is manufactured electronically inside the End Office. Since the called party has not yet picked up the phone, there is no way that the actual audible ring could be transmitted back to the caller. When the called party picks up the handset: EO A detects the off-hook; ringing and ring back will cease; conversation is possible. When the call is finished, the electrical connection that was established within EO A between 5000 and 4422 is broken (hence the term *switched call*). Note that all activity associated with this call took place within the confines of EO A.

Now suppose subscriber 5000 on EO A wants to call a friend at 1100 on EO C. The subscriber dials 678-1100. EO A recognizes that the prefix belongs to EO C and establishes an electrical connection between the caller and a trunk to EO C. This process of prefix interpretation and trunk selection is called *routing*. EO A must now depend on EO C to establish the final connection between the selected trunk and the called party. To enable EO C to do this, the last four digits of the called number must be transmitted down the trunk from EO A to EO C. This process is called *interoffice signaling*. There are numerous signaling schemes, both analog and digital, in common use by telephone operating companies. A full discussion of this subject is far beyond the scope of this book. In any case, when the call is answered we have an electrical connection between 456-5000 and 678-1100 which is carried via the trunk between EO A and EO C.

Before this latest call is broken, imagine that 456-5678 (EO A) calls 678-2457 (EO C). We must repeat the procedure; therefore we need another trunk between EO A and EO C. Fortunately, one is available. The second call will be connected, and both trunks between EO A and EO C are now in use. Suppose 456-4422 (EO A) now wants to call 678-9110 (EO C). This call cannot be connected even if the called party is not currently using the phone. The caller will hear something that sounds like a busy signal (the same audio tones are used), but it will pulse at twice the rate of a normal busy signal. There are also cases when a call may not be completed because there are an insufficient number of connecting links available inside the End Office. The caller will receive a similar tone signal. These problems are called *congestion* or *blocking*.

What we have here is a problem of resource allocation. The system in Figure 7–4 makes it possible for any subscriber to contact any other subscriber, but they cannot all call at the same time. On the other hand, if we provided enough trunks to totally prevent congestion problems, we would have a very uneconomical network because most of the trunks would be idle most of the time. Telephone companies are constantly adjusting their networks to make the available facilities match the demand for connections. Unfortunately, the demand is irregular and there will always be times, such as holidays, when the system will be overloaded.

The three End Offices in Figure 7–4 have trunks to other towns as well. If we wish to provide every phone in the nation with access to every other phone, must we also have trunks linking every small town to every possible call destination? Once again, this is obviously expensive and impractical. What we do have is the situation illustrated in Figure 7–5. There are clusters of End Offices (A,B,C; D,E) with trunks to a nearby medium sized city (M1). A subscriber on EO A who wishes to call a subscriber on EO D must be routed through a higher level office (M1). For a call from EO A to EO F, the routing would be from A, through M1, to the closest large city (L1), down to M2, and finally to EO F. This is called a *hierarchical network*. In the United States the hierarchy extends from

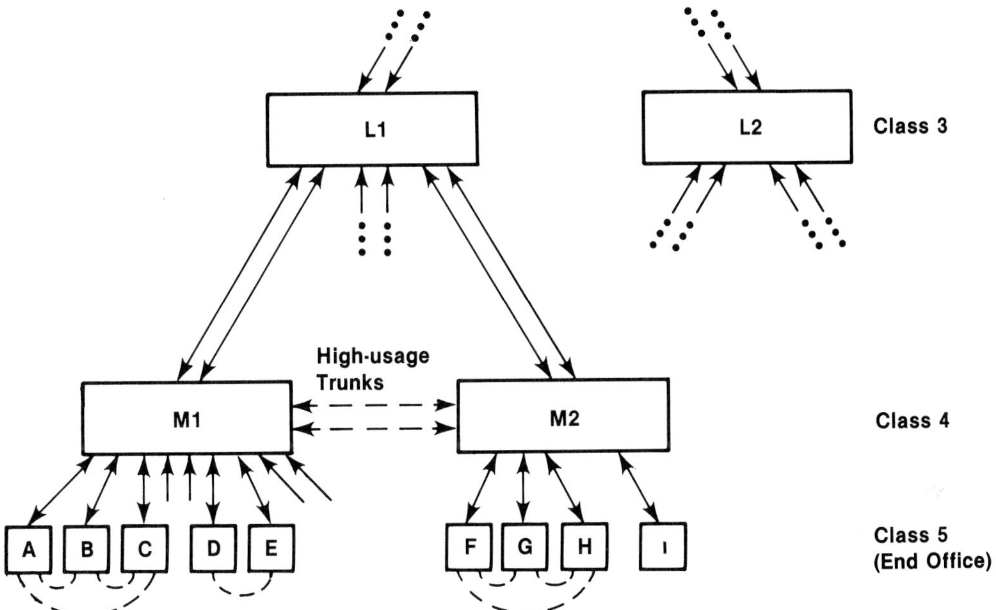

FIGURE 7–5   Long-Distance Telephone Switching

local end offices (also called *class 5 offices*) up to regional centers (called *class 1 offices*). The levels and their formal names are

CLASS	OFFICE TYPE
1	Regional Center
2	Sectional Center
3	Primary Center
4	Toll Center
4X	Intermediate Point
5	End Office

There are some exceptions to the pure hierarchical structure. For example, if there is a large volume of calls between two class 4 offices, such as M1 and M2, *high-usage trunks* (shown by dashed lines in Figure 7–5) might be installed to avoid the trip up to L1 and back.

As we go to higher levels in the network, the physical nature of the trunks changes. They may be carried by microwave links, fiber-optic cables, or even communications satellites, rather than simple wires. At these levels very large numbers of trunks are required (on a link from San Francisco to New York, for example). The media, therefore, use multiplexing techniques to carry thousands of simultaneous calls on a single link.

7·3   THE NATURE OF LONG-DISTANCE CONNECTIONS

The 2-wire subscriber loop (signal and return) is capable of carrying audio signals in both directions. As long as call distances are short, a 2-wire circuit can be used from end to end. However, as we increase the length of the connection, the signal becomes weaker until we get to a point where amplification is a necessity. It is an unfortunate fact that electronic amplifiers have a distinct input and output. That is, they will only amplify signals moving in a single direction. To deal with this problem, phone companies have created a device called a *hybrid*. Its application is illustrated in Figure 7–6.

A hybrid converts a 2-wire 2-directional circuit into a 4-wire 2-directional circuit. On the 4-wire circuit, a pair of wires is dedicated to each direction of transmission. Audio energy from West subscriber is routed by its hybrid to the input of the eastbound *amplifier* (ignore the *echo suppressor* for the moment). Only a small fraction of the energy coming from West subscriber leaks over to the other half of the 4-wire circuit (toward the output of the westbound amplifier). Since this small amount of leakage will not propagate backwards through the westbound amplifier, it causes no problem.

We now have a good, strong signal out of the eastbound amplifier. It enters the East hybrid, which routes most of the energy to the 2-wire circuit to East subscriber. Once again, a small fraction of the energy leaks back onto the westbound section of the 4-wire circuit. In this case we have a problem. The leakage is in the direction that will be amplified by westbound amplifier and will be heard as an echo by West subscriber. The echo delay can be as much as 100 ms, which is extremely annoying to a human speaker, hence, the echo suppressors.

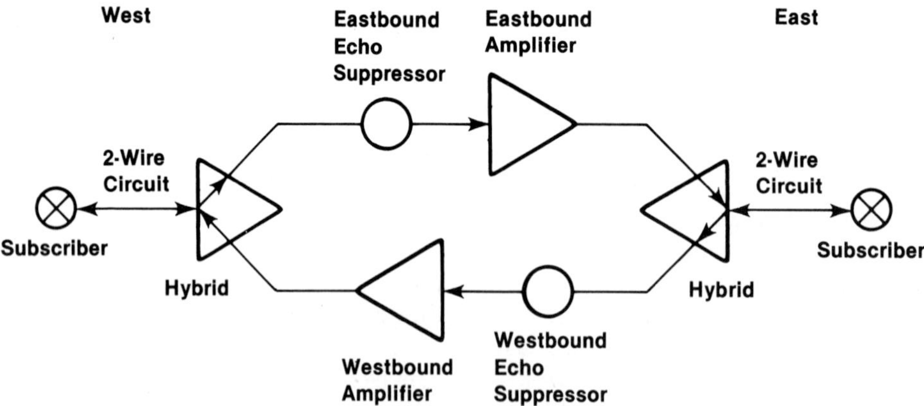

FIGURE 7–6   A Typical Long-Distance Circuit

The telephone network is primarily designed for human conversation, which is (usually) half-duplex with a rather long pause at the time that transmission direction turns around. Telephone operating companies therefore installed circuits which detect the direction of energy flow and turn on the echo suppressor in the opposite leg of the 4-wire circuit. The speaker is thus spared the annoyance of hearing their own echo. Since there are usually short pauses in speech between words and sentences, the echo suppressors are designed to remain on until there is at least 400 ms of silence on the line. The turnaround time of human conversations is usually longer than 400 ms, and speakers are totally unaware of the presence of the echo suppressors.

While this is all well and good for human conversations, it is horrible for data communications. It makes full-duplex communications impossible. For the case of half-duplex communications, the turnaround times become a very significant fraction (perhaps more than 50%) of the total length of the call. Fortunately, the telephone operating companies have produced a solution which satisfies both voice and data transmission needs. A sustained tone (at least 400 ms) in the frequency range from 2010 Hz to 2240 Hz will disable the echo suppressors. In many modems the frequency of the answer tone is chosen to be in this range. As long as there is continuous transmission of carrier energy in either direction, the echo suppressors will remain disabled. For full-duplex modems there is obviously a constant transmission of carrier. For half-duplex modems the carrier direction can be reversed so rapidly that the echo suppressors never notice and, therefore, remain disabled for the length of the call.

This means that data communications equipment must be designed to work in the presence of echoes. Full-duplex modems designed to operate on 2-wire circuits use different carrier frequencies for each direction. The modem receiver can therefore employ electronic filtering to "tune out" the echo of its transmitter. Since half-duplex modems typically use the same carrier frequency for both directions, we must use the RTS–CTS delay described in the previous chapter. The 150 ms we have to wait for the echoes to fade away is still much better than the 400–500 ms required for the echo suppressor to turn off if it could not be disabled by the answer tone.

In this entire section we have been talking about switched circuits. (We conveniently ignored the electrical connections established by the various offices the call passed through, but these have no effect on the subject of echo suppression.) Suppose, however, we want a dedicated circuit. We could lease a 2-wire dedicated circuit, but, if it was long enough, it would really look just like Figure 7–6. Since the hybrids are usually located in the subscriber's local End Office, most of our 2-wire circuit is, in fact, a 4-wire circuit. Realizing this, we could ask the phone company to eliminate the hybrids and echo suppressors entirely and extend the 4-wire circuit all the way to the ends of our dedicated connection. Telephone operating companies do offer this type of connection, and, for long distances, the cost of an end-to-end, 4-wire dedicated circuit is only slightly higher than a 2-wire dedicated circuit. Since we now have a separate pair of wires for each carrier direction, we can get full-duplex operation from modems that normally would be half-duplex on a 2-wire circuit. Even if the DTEs using this circuit cannot operate full-duplex, the elimination of turnaround delay results in a large improvement in the speed of message transmission.

7·4   THE HISTORY OF A DATA CALL

We have now accumulated enough background information to follow the history of data calls in detail. The equipment involved in our first example is shown in Figure 7–7, and the timing of significant events is shown in Figure 7–8. At the West side of our diagrams we have a host computer communications port connected to a modem with auto-answer capability. Perhaps this host computer is a data bank containing stock and bond prices, football scores, ski conditions, and other interesting information. A number of these systems have become available to the public in recent years. At the East side we have the owner of a personal computer (with terminal emulation software) who wants to know the price of AT&T common shares. This person also owns a modem and a telephone which, in addition to all the usual components, has a *TALK/DATA* switch. This switch can transfer the subscriber loop wires from the telephone to the modem. This, of course, is only one of many possible equipment config-

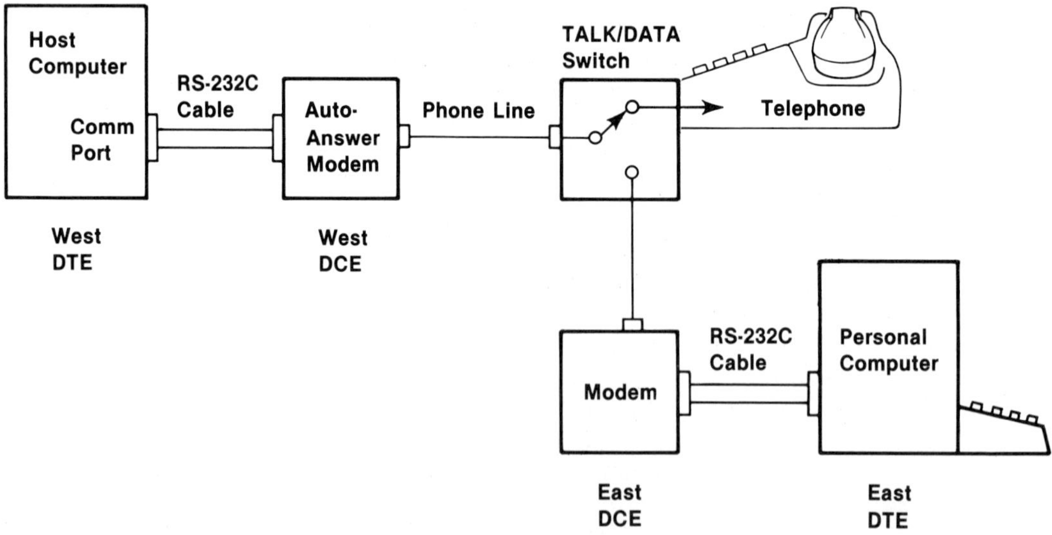

FIGURE 7–7   Equipment Required For a Manually Originated Data Call

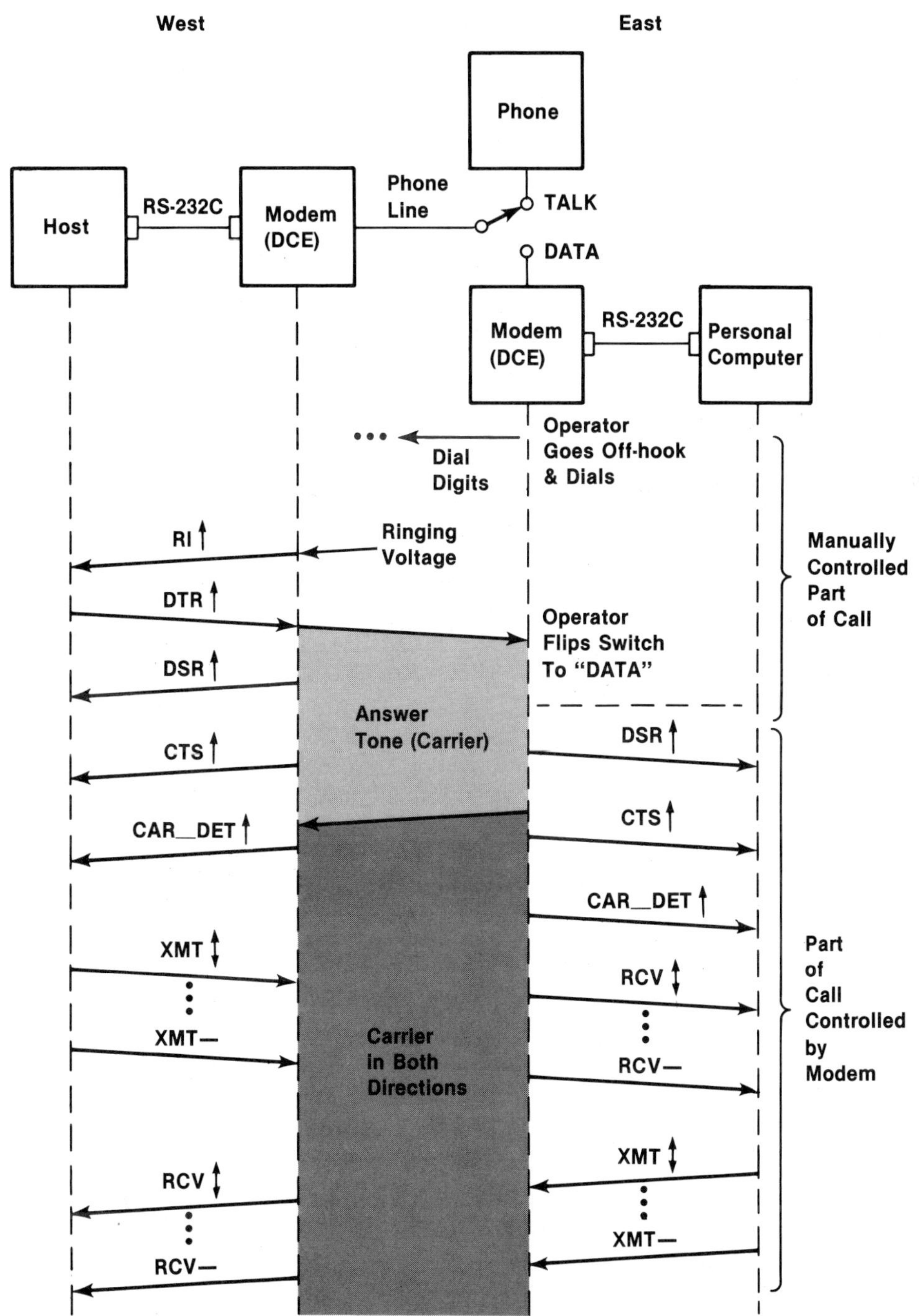

FIGURE 7–8   The History of a Data Call

urations. The procedure we are about to describe is typical, but there are many variations. We make the following assumptions for our example:

- The call is manually dialed from East.
- It is automatically answered at West.
- The modems are asynchronous and can operate full-duplex on a 2-wire switched line.

We start with the switch in the TALK position (phone line connected to the telephone). The PC owner goes off-hook, listens for dial tone, and dials the number of the data bank. The telephone company performs the service of converting dial digits at the East side of Figure 7–8 into a ringing voltage at the West side. Our data bank host computer is prepared to answer and, therefore, asserts DATA TERMINAL READY. West DCE will go off-hook, begin to transmit answer tone, and turn on DATA SET READY. For the simple full-duplex asynchronous modems we have assumed, the answertone is actually the carrier frequency (which is chosen to be in the range that will disable the echo suppressors). In fact, since no carrier control is required for full-duplex operation, the REQUEST TO SEND wire may not even be present in the RS-232C cable. West DCE will simply turn on CLEAR TO SEND shortly after DATA SET READY.

At the East end, the PC owner is still holding the telephone handset. The PC owner hears the answer tone from West and therefore knows that the connection is established. The PC owner immediately flips the switch to DATA position, transferring the wires of the phone line to the DCE. This is a critical point in the progress of our call. Since East DCE does not have the ability to originate a call, this operation must be done from some other device. Once the connection is established East DCE must quickly be given control of the line. At West DCE, the purpose of the delay between DATA SET READY and CLEAR TO SEND is to allow the TALK/DATA switch operator some reaction time before data transmission starts.

Now that the switch is in DATA position, East DCE senses the incoming answer tone. East DCE will therefore begin transmitting its own carrier, turn on DATA SET READY, and CLEAR TO SEND. We have now established connection and have active carriers in both directions. The PC owner can sit down at the keyboard and seek information from the data bank. The first transmission will be originated by the data bank, and it will most likely be a message such as

XYZ INFO BANK
PLEASE LOGON:

The PC owner will have to enter an identification code (so the data bank operator can bill the PC owner for services provided). After the login is complete, the PC owner can enter inquiries, and the data bank will respond. In spite of the full-duplex capabilities of the equipment, the usage is half-duplex. (However, West DTE may be providing a remote echo function for the PC owner's keystrokes.)

The "conversation" proceeds until the user is satisfied and wants to conclude the session. The user will then enter a logoff message. This will prompt West host computer to do a billing computation and then initiate a disconnection sequence (not shown in Figure 7–8). The disconnect will take place more or less as described in Chapter 6. Both modems (if their options have been properly selected) will go on-hook. The PC owner must see that

- the telephone is hung up;
- the switch is returned to the TALK position.

A number of devices have been designed to accomplish both these functions in a single step. One such device, the *exclusion key*, is illustrated in Figure 7–9.

The exclusion key is a modified version of the switchhook "button" that pops up when a telephone handset is lifted from its cradle. Although this phone looks ordinary, it is specially designed to operate in conjunction with a modem. When the handset is lifted, the exclusion key places the phone off-hook and allows dialing in the normal manner. When an answer tone is heard

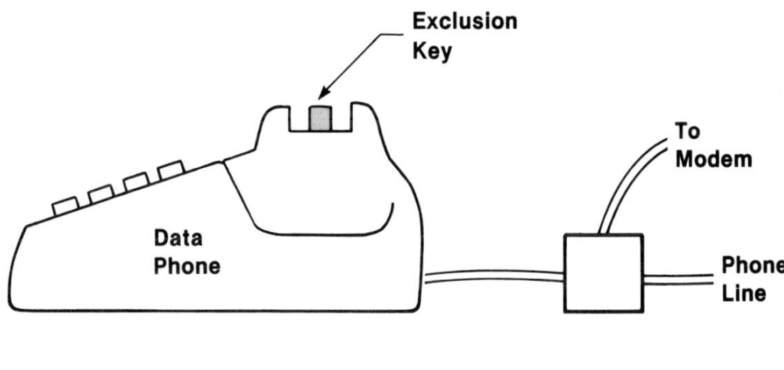

(A) Off-hook Position (TALK)

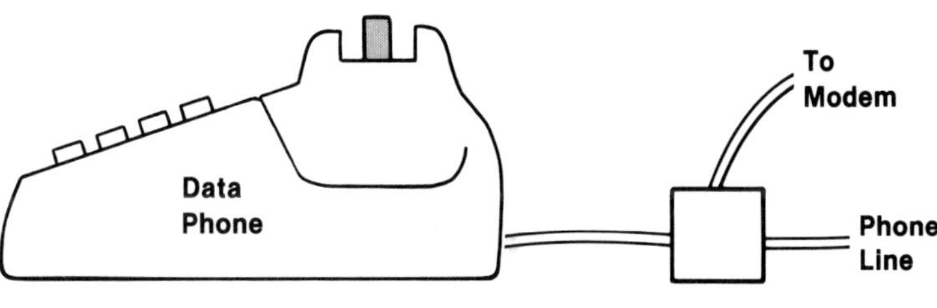

(B) Off-hook Position (DATA)

FIGURE 7–9   Exclusion Key Operation

the operator lifts the exclusion key to a slightly higher position. This switches the line from the TALK position to the DATA position. The handset is left out of the cradle for the duration of the call. When the handset is replaced, the telephone switchhook contacts are opened (on-hook condition), and the TALK/DATA switch is returned to the TALK position.

7·5 AUTOMATIC DIALING

There is no reason why the manual dialing procedure described in the last section cannot be imitated by an electronic device, and, as you might imagine, such devices have been built. They are known as *Automatic Calling Units* (ACU) and are used in configurations such as the one illustrated in Figure 7–10.

The ACU operates a switch which is functionally identical to the TALK/DATA switch that was manually operated in the last section. The ACU must obviously be controlled by a program in the host computer (DTE). The DTE/ACU interface is standardized according to the RS-366 specification published by the Electronic Industries Association (the folks who brought you RS-232C). RS-366 procedures are quite simple and closely parallel the way a human operator uses a telephone:

1. The DTE indicates its desire to initiate a call by turning on a signal to the ACU called *CALL REQUEST*.
2. The ACU takes the subscriber loop off-hook and, when it detects dial tone, turns on a signal to the DTE called *PRESENT DIGIT*.

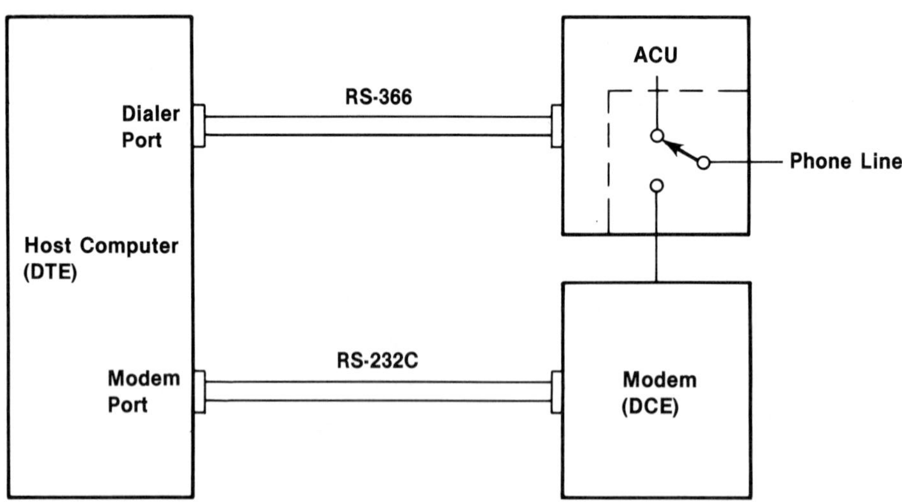

FIGURE 7–10   Auto Call Unit Configuration

3. The DTE responds by presenting a dial digit to the ACU. Digits are presented in binary coded decimal format on a 4-wire parallel bus (RS-366 signals NB1–NB4).
4. The ACU accepts the digit, "dials" it, and then turns on PRESENT DIGIT again. (ACUs may be designed for either pulse or DTMF dialing.)
5. Steps 3 and 4 are repeated until the entire number is dialed. The next time the ACU asserts PRESENT DIGIT, the DTE responds with an *end-of-number* code on the NB1–NB4 bus. The decimal value of this code is 12, so it cannot be mistaken for a digit. The use of end-of-number allows variable-length phone numbers to be dialed.
6. When end-of-number is received the ACU starts a "timeout" and listens for answer tone. The duration of the timeout is user selectable and can be set between limits of 10–40 s.
7. If the timeout expires before answer tone is received, the ACU turns on a signal to the DTE called ABANDON CALL AND RETRY. The DTE should respond by turning off CALL REQUEST. The ACU then places the line back on-hook.
8. If an answer tone is received before the timeout expires, the ACU flips the switch to the DATA position, exactly as the human operator did in the previous section. We now have a modem to modem connection controlled by the RS-232C interface.

The RS-366 device gives us the capability of automatically dialing calls at the expense of adding another external device to our collection of peripherals and another connector to our host computer. For this reason freestanding ACUs are rapidly being replaced by devices called *smart modems*. This is particularly true in personal-computer applications.

What makes an ordinary modem "smart"?—the addition of a microprocessor chip and some firmware. The primary function of the firmware is to incorporate dialing functions into the modem. We have now replaced our ACU/modem combination with a single device connected to our DTE by one RS-232C cable. We have also eliminated the need for the TALK/DATA switch.

By now we are intimately familiar with the functions of all the wires in the RS-232C interface, and it is obvious that none of them are suited for controlling a dialing device. The only way to get dialing instructions from our DTE to the smart modem is by sending ASCII characters down the RS-232C TRANSMIT DATA wire. This raises yet another question. How is this modem smart enough to know if an ASCII character is part of a dialing instruction or if it is merely data to be transmitted on the phone line? The smart modem solves this problem by operating in two different states:

- the off-line state in which data characters from the DTE are interpreted as modem commands
- the on-line state during which we have conventional modem operation (data from the DTE is used to modulate the transmitted carrier)

When a smart modem is initially turned on, it enters the off-line state. The commands it will accept fall into three categories:

- dial a number
- answer an incoming call
- change one of the smart modem's internal option settings

The first two categories will usually result in the smart modem going to the on-line state. There are a number of ways it can be returned to the off-line state:

- no answer in response to a dialed number
- loss of carrier during a connected call
- an *escape code* from the DTE

The escape code allows the smart modem to go off-line while a call remains connected. The escape code must therefore be a character sequence that is unlikely to occur within a normal message. A typical sequence is a silent interval of at least one second followed by + + +. Once back in the command mode the smart modem can be instructed to do a number of interesting things, such as sending DTMF signals to computer systems that respond to these tones (there are some systems that allow users to check their bank balances or pay bills in this manner). Other commands place the line back on-hook or return to the on-line state.

The RS-232C signals that control the smart modem can be transmitted by software, or, in the case of a personal computer running as a terminal emulator, the commands can be issued directly from the keyboard. Whoever (or whatever) is controlling the smart modem must be constantly aware of the on/off-line state of the modem. The smart modem has reduced hardware costs by eliminating the ACU, but it has increased the complexity of controlling the RS-232C interface.

Some other interesting features that may be provided by smart modems are

- redial the last number;
- set up a dialing directory which allows frequently called numbers to be selected from a menu on the terminal's screen;
- select pulse or DTMF dialing;
- turn on a small speaker that allows the user to hear call-progress signals such as ring back or busy tones;
- enable auto-answering (even ordinary modems can do this).

REVIEW QUESTIONS

1. If the time between digits is approximately 1 s, how long does it take to dial (800) 999-7654 on a rotary phone?

2. What frequencies will be generated when you dial 911 on a DTMF telephone?

3. Referring to Figure 7–4, list the significant events that take place at EO B when 567-1234 calls 456-5000. Assume that trunks are available and the called party is not busy.

4. Draw a diagram of a call that has to be routed through a class 2 office to establish a connection between two subscribers. How many links (trunks plus subscriber loops) are involved in the call?

5. Explain why full-duplex communications would be impossible on a long-distance switched circuit if echo suppressors could not be disabled.

6. Describe the purpose of the TALK/DATA switch on telephones operated in conjunction with a modem.

7. What is the function of the end-of-number code on the RS-366 ACU interface?

8. What would happen if an RS-366 ACU attempted to call a DTE which was holding DATA TERMINAL READY continuously in the off state?

chapter eight

IMPROVED DTE/DCE INTERFACE CIRCUITS

In recent years the Electronic Industries Association and other standards agencies have published new specifications for DTE/DCE interface circuits. These specifications attempt to deal with some limitations and deficiencies present in RS-232C devices. In spite of these improvements, the data communications community has not rushed to embrace these new standards, and the vast majority of modems and terminals are still manufactured with RS-232C connectors. We will briefly look at these new standards. The primary purposes will be to learn what the limitations of RS-232C are; how they can be circumvented; and why these improvements have not generated a great deal of enthusiasm among a group of people who are usually very quick to adopt any "advance in the state of the art."

8·1   THE LIMITATIONS OF RS-232C

There are three factors that limit the application of the RS-232C interface:

- distance
- data rate
- noise

Distance

In the multiwire cables, such as those used to carry RS-232C signals, there is a capacitance between any signal wire and the common return wire (SIGNAL GROUND). The RS-232C specification states that the capacitance seen by a driver circuit shall not exceed 2500 pF. To comply with the letter of the RS-232C document, a driver can be designed to operate with a capacitive load of exactly 2500 pF and nothing greater. The capacitance increases in proportion to cable length, and 2500 pF corresponds to approximately 50 ft (15 m) of typical cable. Most RS-232C drivers have capabilities that exceed the limits of the specification by a wide margin, and, in practice, RS-232C interfaces operate reliably on cables much longer than 50 ft. This is particularly true for low baud rates.

Data Rate

The RS-232C specification explicitly states, on the first page of text, that it does not apply to signal rates greater than 20 000 bits/s. Once again, the problem is related to cable capacitance. RS-232C specifies a fairly high input resistance for the receiver circuit (between 3000 ohms and 7000 ohms). Without going deeply into transmission line theory, we can say that this resistance limits the rate at which the cable capacitance can be charged and discharged. This leads to the rounding of signal edges and the distortions that we discussed in Chapter 2. At high data rates, the rounded "leading edge" becomes a significant portion of the total bit time.

Again, this limitation seems to make little difference in practice because the fastest telephone modems available operate at 19 200 bits/s, and the vast majority at much lower rates. It is possible to lease special "wideband" lines which permit operation at much higher baud rates, and RS-232C circuits obviously cannot be used with these. In later chapters we will also see an interface to a local area network that requires very high data rates, and, here too, it will not be possible to use RS-232C drivers and receivers.

Noise

Noise can distort and change the value of a binary signal on a cable. There are two possible noise sources:

1. External. Spurious signals which are picked up from the environment (electric motors, radio transmitters, fluorescent lights).
2. Internal. If the common signal return (which carries the combined current of all wires in the RS-232C cable) is not a perfect conductor (0 ohms), the current contributed by one wire causes a voltage drop in the common return wire that is sensed by the receivers for all other wires.

Both of these problems are exaggerated by cable length. Long cables make better "antennas" for pickup and obviously have a higher resistance. Once again, this seems to make little difference in practice. RS-232C circuits have a degree of noise immunity because of the transition region between the positive and negative states (see Figure 6–2). External noise problems can usually be eliminated by using shielded cables, and the wire gauges used in typical RS-232C cables are heavy enough to make voltage drops in the return wire insignificant (even for cables much longer than 50 ft).

Nevertheless, the new standards provide more sophisticated solutions to these problems. Most of these benefits derive from the use of *balanced interface circuits*. The RS-232C interface, with one wire for each signal and a shared return wire, is an *unbalanced interface circuit*. Balanced circuits require a 2-wire twisted pair for each signal and a radically different design concept for driver and receiver circuits. We will study balanced circuits in more detail later in this chapter.

8·2 NEW CIRCUITS AND NEW FUNCTIONS (RS-449)

Between 1975 and 1977 the Electronic Industries Association published a family of new specifications for the DTE/DCE interface. There are three documents in this family:

- RS-449 describes the functions of the signals that make up the interface.
- RS-422 describes the electrical characteristics of balanced drivers and receivers.
- RS-423 describes the electrical characteristics of unbalanced drivers and receivers.

The full RS-449 interface is composed of two connectors:

- a 9-pin connector which carries all signals associated with the secondary channel
- a 37-pin connector which carries everything else

If a modem does not have a secondary channel (most do not), the 9-pin connector is not present.

RS-449 divides its signals into two different categories:

- Category I signals may be balanced in some applications and unbalanced in others. In order to accommodate both situations with the same connector design, two pins are assigned to all Category I signals. In equipment that uses these signals in the unbalanced mode, one pin is not connected.
- Category II signals are always unbalanced and therefore have only one connector pin assigned to each of them.

Every signal defined for RS-232C has an equivalent signal in RS-449. Some of the names are slightly different, but their functions are identical. RS-449 also includes a few additional signals which we will get to shortly. The assignment of a signal to Category I or Category II depends on its potential for high-speed operation and noise susceptibility. Signals in Category I (using their RS-232C names) are

- TRANSMIT DATA
- RECEIVE DATA
- REQUEST TO SEND
- CLEAR TO SEND
- DATA SET READY
- DATA TERMINAL READY
- CARRIER DETECT
- TRANSMIT CLOCK (either source)
- RECEIVE CLOCK

If the DTE/DCE interface is operating at 20 000 bits/s or more, Category I circuits must be operated in the balanced (2-pin) mode. Below 20 000 bits/s the choice of balanced or unbalanced mode is at the option of the equipment designer. All signals other than those listed previously (including everything associated with the secondary channel) are assigned to Category II and always operate in the unbalanced (1-pin) mode.

The RS-449 signals without RS-232C equivalents have been introduced to deal with some reliability and maintainability problems that frequently occur in communications networks. Imagine a large network consisting of widely separated devices (modems, terminals, and host computers manufactured by several different vendors) linked by lines that belong to one or more telephone companies. When communications fail, which repair service do you call? There are circuits on the RS-449 interface that help to answer this question and also help to keep the network going while you are waiting for the repair service to arrive. These circuits fall into the following two groups.

Local Loopback, Remote Loopback, Test Mode

When a DTE turns on *local loopback,* the modulator output of its local modem is removed from the phone line and connected back to the demodulator input (see Figure 8–1). If a link to a remote device is inoperable, but the DTE is able to receive echoes of its own transmissions with the DCE in local loopback,

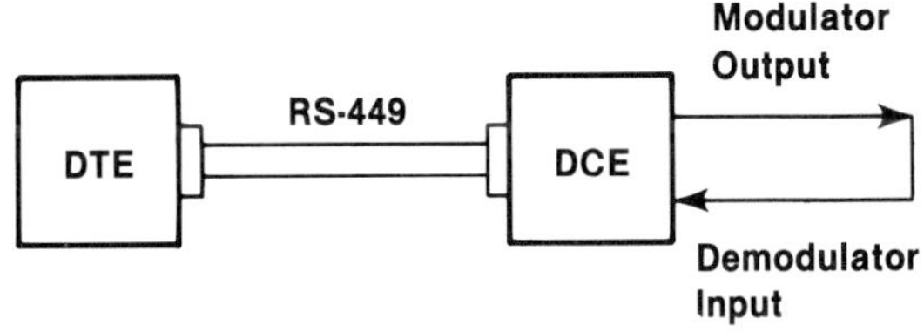

FIGURE 8–1   Local Loopback

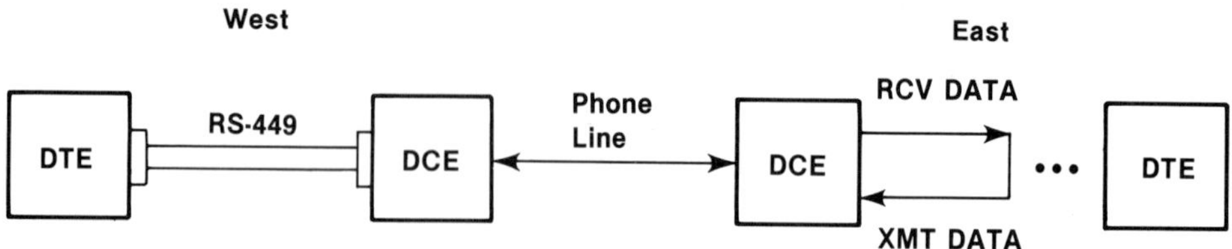

FIGURE 8–2   Remote Loopback

the problem is either in the phone line or in the equipment at the other end. If no echoes are received the problem is in the local DCE or the interface circuits.

If the equipment checks out in the local loopback test, the next step is *remote loopback*. In Figure 8–2 West DTE has turned on remote loopback. This will cause West DCE to send a special signal to East DCE. In response East DCE will connect its RECEIVE DATA output back to its TRANSMIT DATA input. This disconnects East DTE (electrically, not physically) from the link. If West DTE is capable of receiving echoes of its own transmissions in remote loopback mode, the problem lies in East DTE.

The *test mode* signal is turned on by the DCE to indicate to the DTE that real communication is impossible because a loopback mode has been activated. For example, in remote loopback both DTEs will have test mode turned on.

Select Standby, Standby Indicator

Select standby is a signal which may be turned on by the DTE to instruct the DCE to switch from its normal communications medium to some alternative. A modem which normally operates on a dedicated circuit may transfer over to a switched line when the dedicated circuit fails. The DCE must obviously be designed to accommodate this feature. A transfer such as the one described previously may involve a change from full-duplex to half-duplex operation and a reduced baud rate.

Standby indicator is the DCE's response to select standby. It is turned on to indicate to the DTE that the requested transfer is complete.

8·3   UNBALANCED DRIVERS AND RECEIVERS (RS-423)

RS-423 defines the electrical characteristics of unbalanced interface circuits. Schematically these appear virtually identical to RS-232C drivers and receivers (see Figure 8–3 and compare it to Figure 6–3). The "ground" symbols represent the common return wire shared by the balanced circuits in the cable.

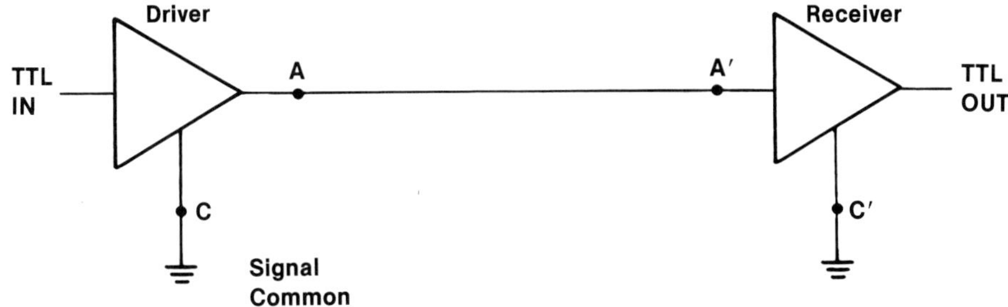

FIGURE 8–3   RS-423 Unbalanced Interface Circuits

In the MARK, or OFF, state, driver output A is negative with respect to C (signal common on the driver side of the cable), and, conversely, in the SPACE, or ON, state A is positive with respect to C. The receiver measures its input (A') with respect to signal common at its own end of the cable (C') and produces a TTL output in accordance with Table 8–1.

TABLE 8–1   Unbalanced Receiver
Characteristics

RECEIVER INPUT VOLTAGE ($V_{A'C'}$)	TTL OUTPUT	
< 0	1	(bit value)
> 0	0	(bit value)

$V_{A'C'}$ is the voltage from input A' to common point C'. (In this section and the next we use the notation V_{XY} to indicate a DC voltage measured from point X to point Y with the voltmeter connected so as to give a positive reading if X is positive with respect to Y. If the meter leads are reversed, we get V_{YX}, which is equal to $-V_{XY}$.)

Electrically, the voltage levels and impedances are sufficiently similar to those of RS-232C to allow what the specification calls "interoperation." That is, connecting RS-232C drivers to RS-423 receivers and vice versa. As long as certain precautions, such as overvoltage protection are incorporated, these connections will operate reliably. The DTE/DCE cable will have a 37-pin connector on one end and a 25-pin connector on the other.

There are, however, some important differences between the electrical characteristics of RS-423 and RS-232C circuits. Earlier in this chapter we stated that the cable capacitance problem that arose in RS-232C was related to the relatively high input resistance of the receiver circuit (3000–7000 ohms). RS-423 drivers are designed to handle heavier current loads and typically operate with receiver input resistances in the area of 400 ohms. This reduces both the signal distortion effects of the cable capacitance and the tendency to pick up noise from external sources. Even in the unbalanced mode RS-449 circuits can therefore operate at greater cable lengths and higher data rates than their RS-232C relatives. The following table summarizes recommended maximum cable lengths for RS-423 circuits as a function of data rate.

DATA RATE (BITS/S)	MAXIMUM CABLE LENGTH (FT)
< 900	4000
10 000	400
100 000	40

8·4 BALANCED DRIVERS AND RECEIVERS (RS-422)

RS-422 describes the electrical characteristics of balanced interface circuits. It is apparent from Figure 8–4 that these are different from anything we have thus far encountered. The driver generates a *differential signal* at outputs A–B, and a twisted pair of wires is typically used to connect the driver outputs to the receiver inputs (A'–B'). In the MARK, or OFF, state the output voltage at A is negative relative to the output voltage at B, and, conversely, in the SPACE, or ON, state A is positive relative to B. Note that while the voltage at each output can be measured with respect to its local ground (C), the binary states are defined in terms of the difference between these voltages ($V_{AC} - V_{BC}$).

The receiver measures the voltage of each of its inputs (A' and B') relative to local ground (C') and performs an analog computation of the difference, $V_{A'C'} - V_{B'C'}$, between these voltages. The TTL binary output of the receiver is set on the basis of this difference as summarized in Table 8–2.

Compare this to Table 8–1 for unbalanced receivers. This differential signal characteristic produces a tremendous advantage in the ability of balanced circuits to tolerate noise as compared to unbalanced circuits. Balanced circuits exhibit a property called *common mode noise rejection,* which will be illustrated in the following example.

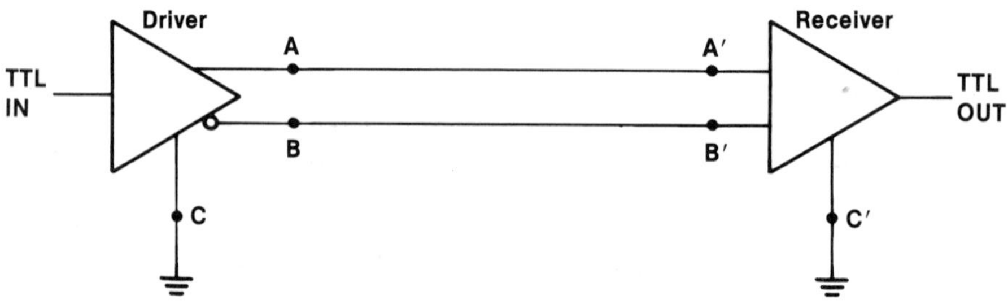

FIGURE 8–4   RS-422 Balanced Interface Circuits

TABLE 8–2 Balanced Receiver Characteristics

RECEIVER INPUT VOLTAGE $(V_{A'C'} - V_{B'C'})$	TTL OUTPUT
< 0	1 (bit value)
> 0	0 (bit value)

Noise induced onto cables (from both the internal and external sources described in Section 8·1) appears as a voltage difference between the ground points at either end of the cable. We will call this noise voltage $V_{CC'}$. Figure 8–5 illustrates the effect of common mode noise on an unbalanced circuit. From basic circuit laws we know that voltages around a closed loop must add up to zero. Starting with our imaginary voltmeter connected between the ground points and proceeding clockwise, we get

$$V_{CC'} + V_{AC} + V_{C'A'} = 0$$

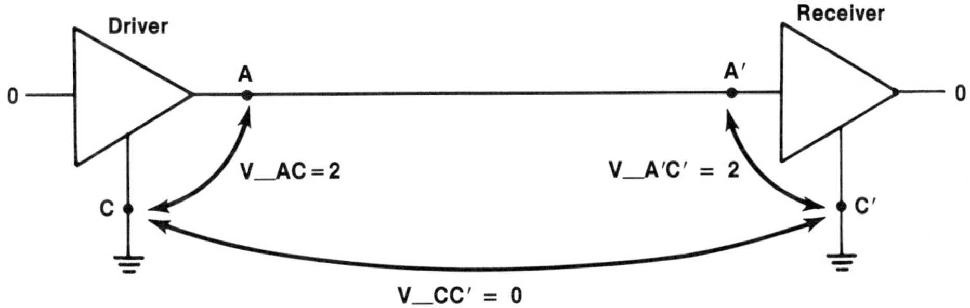

(A) Noiseless Condition

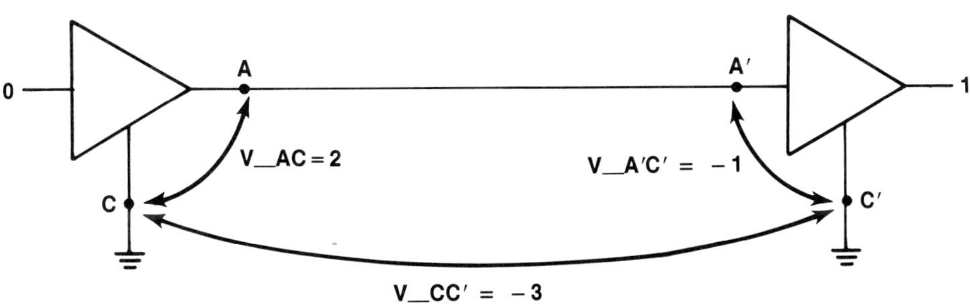

(B) − 3 V Noise Pulse

FIGURE 8–5   Common Mode Noise in Unbalanced Circuits

(As we proceed around the loop, for each measurement the negative lead of the voltmeter must be connected to the point where the positive lead was connected on the previous measurement.)

Since $V_{C'A'} = -V_{A'C'}$ (the receiver input voltage), we can substitute this into the last equation:

$$V_{CC'} + V_{AC} - V_{A'C'} = 0$$

Hence,

$$V_{A'C'} = V_{CC'} + V_{AC}$$

For the noiseless condition (Figure 8–5A) the 0 bit input to the driver produces $V_{AC} = 2$ (driver circuits generate signals relative to their local ground). Since there is no noise, $V_{CC'} = 0$, and we have

$$V_{A'C'} = 0 + 2 = 2\,V$$

According to Table 8–1, the receiver output will be a TTL 0 bit which, as it should, corresponds to the driver input. Now suppose we have a -3-V noise pulse (Figure 8–5B). The driver is still generating $V_{AC} = 2$ in response to the TTL 0 input. We therefore have

$$V_{A'C'} = -3 + 2 = -1\,V$$

producing a TTL 1 at the receiver output, which is obviously an error.

In Figure 8–6 we apply the same -3-V noise pulse to a balanced configuration with similar driver output levels. As we did for the unbalanced case we can compute the receiver input voltages in terms of their respective driver outputs and the ground differential:

$$V_{A'C'} = V_{CC'} + V_{AC}$$
$$V_{B'C'} = V_{CC'} + V_{BC}$$

With the TTL 0 input the driver outputs are

$$V_{AC} = +2\,V$$
$$V_{BC} = -2\,V$$

With no noise,

$$V_{A'C'} = 0 + (+2) = +2\,V$$
$$V_{B'C'} = 0 + (-2) = -2\,V$$

Hence,

$$V_{A'C'} - V_{B'C'} = (+2) - (-2) = 4\,V$$

and according to Table 8–2 the receiver output is TTL 0, which is correct.

If we now apply the same −3-V noise pulse that produced the error in the unbalanced circuits:

$$V_{A'C'} = (-3) + (+2) = -1 \text{ V}$$
$$V_{B'C'} = (-3) + (-2) = -5 \text{ V}$$

and

$$V_{A'C'} - V_{B'C'} = (-1) - (-5) = 4 \text{ V}$$

This still produces the correct TTL 0 output in spite of the fact that the noise drove both receiver inputs below local ground level. Since the receiver responds only to differences in its inputs, it can tolerate common mode noise pulses much greater than those shown. In fact, we would have obtained the same result for the balanced receiver regardless of the size of the noise pulse or the voltage levels generated by the driver (see Review Question 8–4). The only real limitation is the breakdown voltage of the receiver inputs.

What we have just presented is actually an oversimplified analysis used to illustrate the operation of common mode noise rejection. In fact, transition regions (as shown in Figure 6–2) and the use of higher driver output voltages can give unbalanced circuits substantially better noise immunity than this example would indicate.

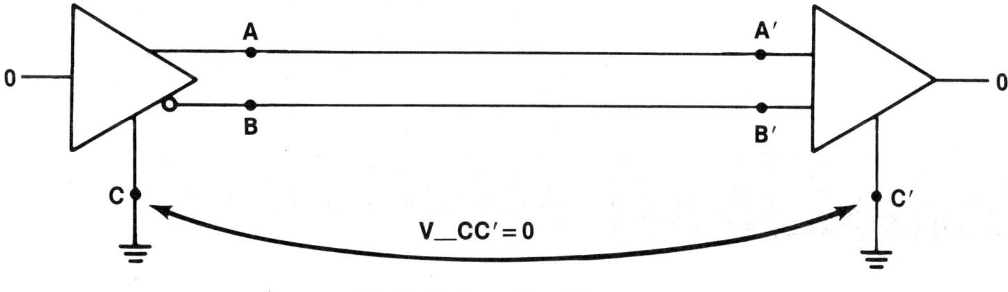

(A) Noiseless Condition

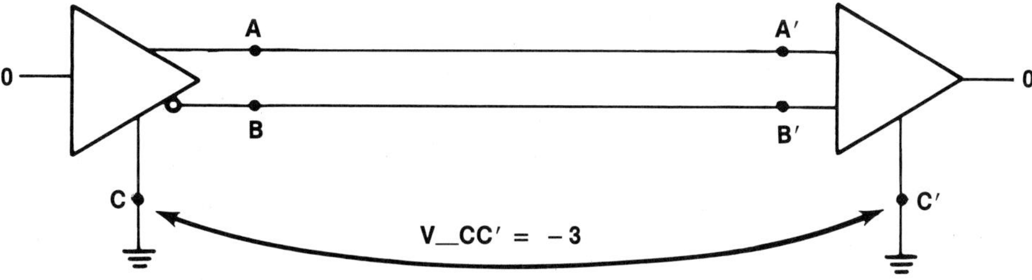

(B) −3 V Noise Pulse

FIGURE 8–6   Common Mode Noise in Balanced Circuits

One final feature that improves the performance of the RS-422 balanced circuits is the 100-ohm resistance that typically appears across the A'–B' receiver inputs. Every electronic transmission system has a *characteristic impedance* (expressed in ohms). For single wires, such as those in an unbalanced interface, this impedance is very difficult to compute and control. For twisted pairs, however, the characteristic impedance is easily determined and, for most cables, is approximately 100 ohms. When a transmission line is terminated with its characteristic impedance, as is the case for RS-422 circuits, the cable capacitance problem disappears. Permissible cable lengths and data rates are therefore very high, as is demonstrated by the following table:

DATA RATE (BITS/S)	MAXIMUM CABLE LENGTH (FT)
< 90 000	4000
1 000 000	400
10 000 000	40

8·5   OTHER APPROACHES (X.20, X.21)

The CCITT (Consultative Committee for International Telephone and Telegraph) is another organization that publishes communications standards. It operates under the auspices of the United Nations and has no connection with the International Telephone and Telegraph Corporation. This organization has published two recommendations for DTE/DCE interface called X.20 and X.21. Both of these interfaces employ a 15-pin connector, making them economical from a hardware point of view. The signals may be implemented as balanced or unbalanced circuits, giving this interface the superior electrical characteristics of RS-449.

As a consequence of the small connector and the option for balanced operation, there are very few signals in these interfaces. X.20 provides only a transmit wire, a receive wire, and some grounds. X.21 adds a clock (making it suitable for synchronous operation) and two control wires. In spite of the sparsity of wires, these interfaces have very powerful DCE control capabilities, including automatic call origination. This is accomplished by sending control characters on the transmit and receive wires which cause numerous state changes in the DCE (similar to the smart modems, but much more complex).

The CCITT has recommended that these standards be adopted for interface to public data networks (we will have something to say about these in a later chapter). In spite of this recommendation very little X.20/X.21 hardware or software has been implemented. Until this happens, the CCITT has recommended the use of its two "interim" standards X.20bis and X.21bis. These are virtually indistinguishable from RS-232C.

REVIEW QUESTIONS

1. Under what conditions might it make sense to use an RS-449 Category I signal in the balanced mode even if the data rate was below 20 000 bits/s?

2. An RS-449 terminal suddenly cannot communicate with its host computer. Where might the trouble lie if the operator found
   a. echoes were returned in local loopback but not in remote loopback?
   b. echoes were returned in both the local and remote loopback tests?

3. In the list of RS-449 Category I signals given in Section 8·2, which ones do you think have the most critical need for operation in the balanced mode?

4. Referring to Figure 8–6 and the equations in Section 8·4, calculate $V_{A'C'}$ and $V_{B'C'}$ for the balanced receiver if the noise pulse was
   a. -8 V
   b. $+10$ V

   (Assume that the receiver outputs remain at $+2$ V and -2 V) Would the receiver output change state in either of these cases? What if the driver outputs were reduced to $+1$ V and -1 V?

5. If the unbalanced receiver in Figure 8–5 had been designed with a transition region from $+1.5$ V to -1.5 V, would the -3-V noise pulse in the illustration cause an output error? Explain why.

chapter nine

MODULATION TECHNIQUES FOR DATA

In the last few chapters we have been casually using the terms "sine wave carrier" and "modulation." If we are to understand what is happening on the analog side of a modem, we must become somewhat more precise about the meaning of these terms. This subject can get highly theoretical, but the intention of this chapter is not to make the reader an expert on modulation theory. Rather, it is to provide an intuitive feel for the operation and design problems of modems.

A sine wave is a mathematical function of time. Suppose the strength of our carrier signal is measured in volts. We have an equation that allows us to compute the signal voltage for all instants of time. This equation is

$$V = A * \sin (2\pi * F * t + PH)$$

If we have a pocket calculator (or even a book of trigonometric tables), we can select values for the time variable t and compute instantaneous values for the voltage V. (Remember that the quantities inside the parentheses are measured in radians, not degrees.) Of course, we could also plot these values on graph paper and get the familiar cyclical waveform.

However, this is not a trigonometry book. The important thing to notice is that there are three adjustable constants (parameters) in our equation that determine the shape of the sine wave. They are

- A: amplitude
- F: frequency
- PH: phase

If these parameters determine what a sine wave looks like, we should be able to view one (on an oscilloscope, for example), make some measurements and compute the values of the parameters. Let us look at a typical oscilloscope trace of a carrier signal (Figure 9–1) and see if this is possible. We can directly

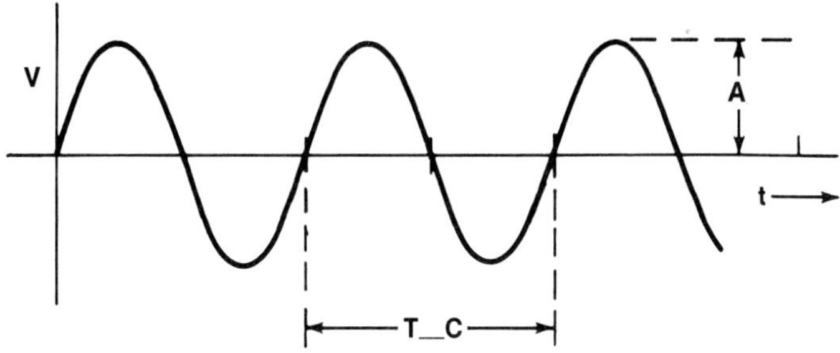

FIGURE 9–1 Oscilloscope Trace of a Carrier Signal

measure the amplitude A from the height of the signal peaks. We can measure the time period of one cycle (T_C) and compute the frequency from the formula

$$F = \frac{1}{T\_C}$$

That takes care of amplitude and frequency. What about phase?

Phase is different. It is meaningless to speak of measuring a carrier signal's phase unless we define a *reference signal*. A reference signal is another sine wave with the same frequency as the carrier. The procedure for measuring phase is illustrated in Figure 9–2. A cycle of the carrier is divided into 360°. The

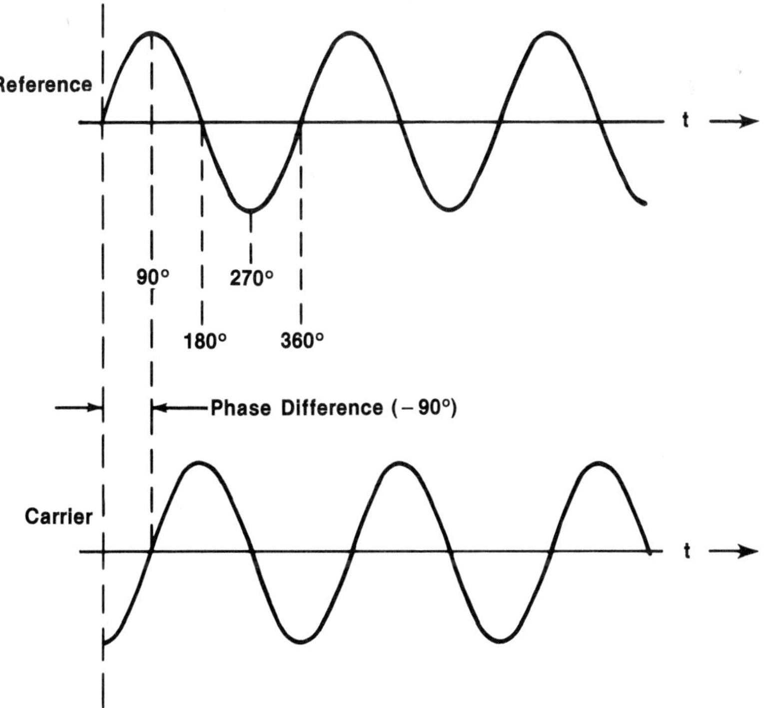

FIGURE 9–2 The Measurement of a Carrier's Phase

phase difference (sometimes called *phase shift*) is the number of degrees between the positive-going zero crossing of the reference and the nearest positive-going zero crossing of the carrier. In Figure 9–2 the carrier is *lagging* the reference by 90°. It is said to be lagging because it arrives at its positive-going zero crossing point at a later time than the reference. Lagging phase shifts are given a negative sign. The phase of the carrier in Figure 9–2 is therefore −90° relative to the reference.

If the positive-going zero crossing of the carrier occurred earlier than that of the reference, we would say that the carrier was *leading* the reference, and the sign of the phase difference would be positive. It is not meaningful to talk about phase differences greater than 360° because it is impossible to tell the difference between a phase difference of D° or (D + 360)° or (D + 720)°, and so on. Because our signals extend indefinitely to the left and right of the small waveform sections shown in Figure 9–2, it should also be obvious that a phase difference of −90° is no different than a phase difference of +270°. Stated mathematically, if

$$\sin(D) = \sin(360 + D)$$

then

$$\sin(-D) = \sin(360 - D)$$

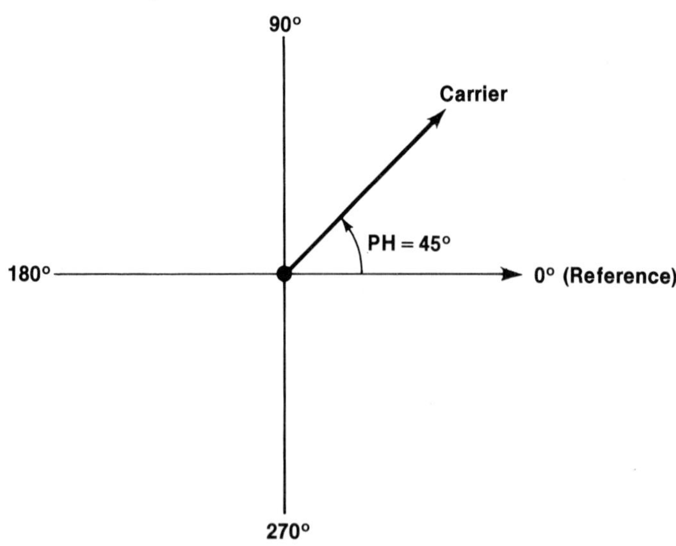

FIGURE 9–3   Phase Vectors in Polar Coordinates

It is inconvenient and difficult to graphically represent carrier phase by constantly drawing pairs of sine waves. There is a simpler method (see Figure 9–3). We draw a polar coordinate system with an arrow pointing horizontally to the right as a representation of the reference signal. The carrier is represented by another arrow (called a *vector*) emanating from the origin of the coordinate system. The phase of the carrier is represented by the angle between the arrows. Positive differences are measured counterclockwise; negative differences are measured clockwise. The 360° maximum and the fact that +45° is exactly the same as −315° should be obvious from this diagram.

9·1 MODULATION

We are now in a position to describe the *modulation* process more precisely:

Select one of the three measurable parameters of a carrier signal (A, F, or PH). Vary the value of the selected parameter in direct relation with the variation of a second signal called the *information signal.*

The opposite process is *demodulation:*

Reconstruct the variation of the information signal by measuring the variations in a parameter of a received carrier signal.

The three techniques at our disposal are therefore

- amplitude modulation (AM)
- frequency modulation (FM)
- phase modulation (PM)

There are also some modulation techniques that involve simultaneous variation of more than one parameter.

Our definitions have made no restrictions on the nature of the information signal, and, in general, it may be a continuously varying analog signal such as voice, music, or the video output of a TV camera. These, however, are not the subject of this book, and we will deal with digital signals only. In Figure 9–4 we have a typical digital waveform as our information signal and examples of how it would modulate a carrier by use of each of the three techniques listed previously. Notice that, in all cases, the carrier frequency must be substantially higher than the rate of variation of the information signal.

The first example is amplitude modulation. We have defined an amplitude value A1 as representative of the SPACE condition of the information signal and A2 as representative of MARK. We apply this rule to our carrier, and the result is Figure 9–4A. Notice how the "envelope" of the carrier follows the digital signal.

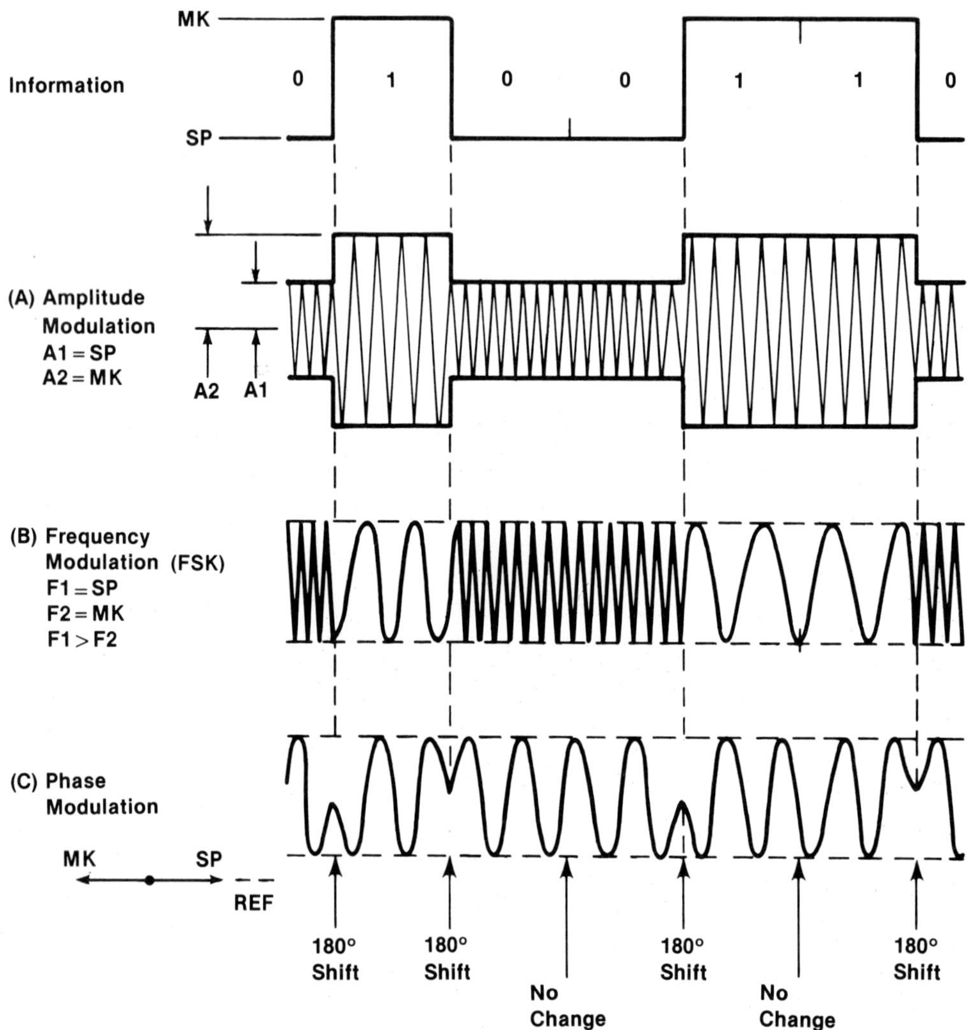

FIGURE 9–4 **Modulation of Digital Signal**

In Figure 9–4B we see a frequency modulated version of the same information signal. We have selected a high frequency to represent SPACE and a low one to represent MARK. There appear to be two different carrier frequencies, but we can think of F1 and F2 as deviations from a single "carrier" halfway between. This binary version of FM is called *frequency shift keying* (FSK).

Figure 9–4C is an example of phase modulation. We have selected MARK and SPACE to be 180° apart, and we therefore have a 180° phase shift whenever there is a transition in the information signal. If you know anything about electronic circuits, you must have guessed that the waveform in Figure 9–4C would be difficult to demodulate. Modems transmit only the modulated carrier and not the reference signal. We therefore have nothing to indicate the phase of the carrier except the occasional glitches shown in Figure 9–4C. Phase modulation techniques used in actual modems must be somewhat more sophisticated than the simple illustration in Figure 9–4C. We will return to this problem shortly.

9·2 MODULATION FOR SYNCHRONOUS DATA

There is one further observation we can make about all the modulation techniques described earlier. At the end of each bit time, if the value of the information signal does not change, there is no detectable change in the modulated carrier. If we transmitted a string of 100 consecutive MARK bits, all of these techniques would produce a pure sine wave for the 100-bit interval. This is exactly what we want for the asynchronous data format. For synchronous data we must make something happen to the carrier at the end of each bit time even if there is no change in the value of the information signal. These regularly spaced events on the carrier allow receiver circuits in synchronous modems to derive the RS-232C RECEIVE CLOCK signal.

A technique suitable for synchronous data is illustrated in Figure 9–5. It is called *differential phase shift keying* (DPSK). It is similar to our earlier example

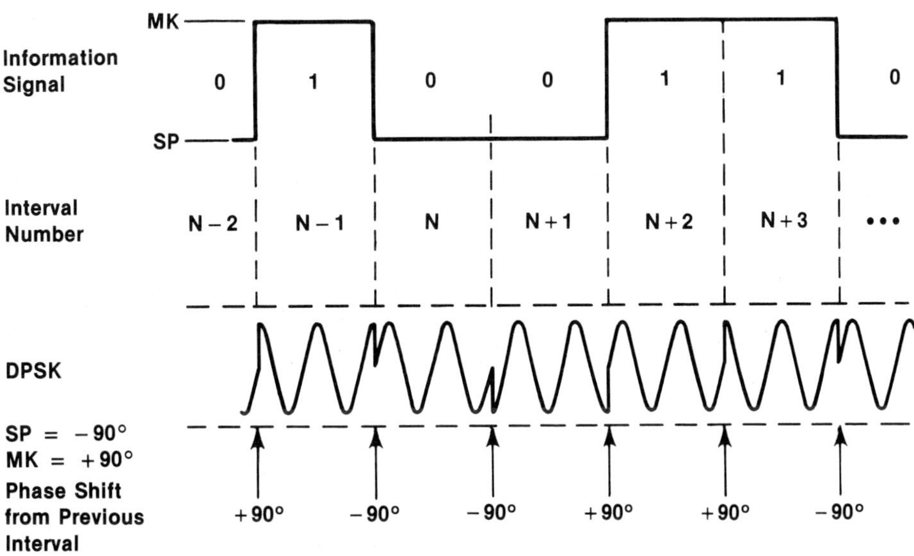

FIGURE 9–5 Differential Phase Shift Keying

of phase modulation in the sense that we have sudden jumps in the phase of the carrier. There are, however, two major differences:

- The phase measurement reference for each bit interval is the carrier in the previous interval
- A phase shift must occur at the end of each bit interval, even if the state of the information signal does not change.

THE ENCODING RULES   At the end of each bit interval, check the state of the next bit in the information signal.

- If the bit is MARK, insert a $+90°$ phase shift.
- If the bit is SPACE, insert a $-90°$ phase shift.

Note that there is no absolute phase reference. Phase jumps are always relative to the current state of the carrier. A $-90°$ phase shift actually means jump the voltage of the carrier back to where it was one-quarter of a cycle ago. Similarly, a $+90°$ phase shift means jump the carrier signal voltage to where it would have been one-quarter of a cycle in the future if no phase shift had occurred. If we symbolically number the bit intervals in Figure 9–5, $N - 1$ is the reference for interval N, N is the reference for interval $N + 1$, and so on. The shift values for each bit transition time are noted at the bottom of Figure 9–5.

A slightly different way of looking at the activity in Figure 9–5 is shown in Figure 9–6. Assume that during bit interval $N - 2$, our DPSK signal is at a phase of 0° relative to some unchanging reference. The first MARK bit (interval $N - 1$) takes the carrier vector to 90°. The two following SPACE bits take us back to 0°

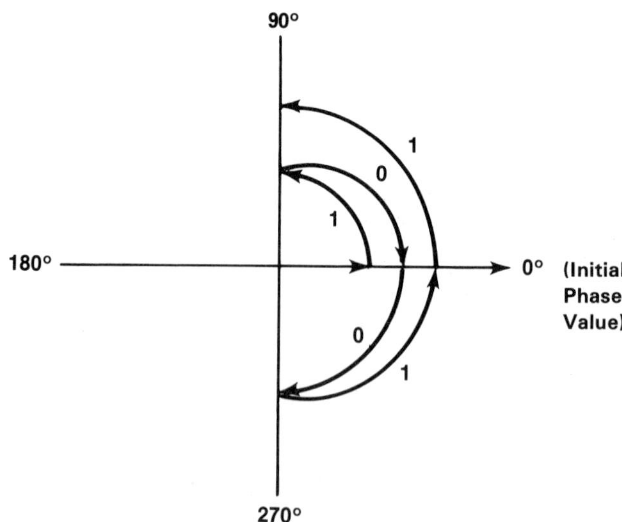

FIGURE 9–6   Polar Plot of DPSK Activity

and then to 270°. Two more MARKs take us counterclockwise back to 90°. A long string of consecutive MARKs would result in a constant counterclockwise rotation of the carrier vector relative to the fixed reference.

The fixed reference is not necessary to demodulate the DPSK signal. Circuits in the modem receiver can lock onto the phase of a bit interval and use this as a reference for measuring the phase of the next interval. Thus, the DPSK technique has provided a solution to two of the design problems we noted in the last few paragraphs:

- We have used the carrier itself to provide a reference for phase measurement.
- The guaranteed occurrence of a phase shift at the end of each bit interval allows modem receiver circuits to derive a clock signal for synchronous data.

9·3 BAUD RATE AND MULTILEVEL MODULATION

All of the modulation techniques we have discussed thus far are digital. Modulation involves the controlled variation of a parameter of a carrier. Digital modulation adds the further restriction that the parameter can be changed only at certain times. When we are discussing a modulated carrier, we shall use the following specialized definition of the term SIGNAL (when it is used in this sense, SIGNAL will always be capitalized):

A SIGNAL is a time interval during which the parameters of a carrier remain constant (that is, the carrier appears as a pure sine wave).

In the modulation examples in Figure 9–4, we get a new SIGNAL each time the information input changes state. A consecutive string of information bits of the same polarity will cause a SIGNAL to be stretched out, but there is a definite *minimum* SIGNAL *time* which, in the examples in Figure 9–4, is the length of one information bit. In the DPSK example in Figure 9–5, all SIGNALS are the same length. We call this length the minimum SIGNAL time for DPSK. Our algebraic symbol for minimum SIGNAL time will be T_SIG.

In Chapter 2 we stated that for binary (two-state) circuits a line's data rate (B_P_S) was equivalent to its baud rate, and we have used these terms interchangeably. We are now prepared to define baud rate more precisely:

Baud rate is the maximum number of SIGNALS a device can transmit in one second.

Since baud rate is a *maximum SIGNAL frequency*, it is obviously equal to the reciprocal of minimum SIGNAL time. Therefore,

$$\text{BAUD\_RATE} = \frac{1}{\text{T\_SIG}}$$

If we assume that the information waveform input in Figure 9–4 represents the RS-232C input to a modem and that the FSK waveform (Figure 9–4B) is the resulting modulated carrier, we can see that the minimum SIGNAL time on the carrier is equal to the RS-232C bit interval (T), which was defined in Chapter 2. Then

$$T = T\_SIG$$

and

$$\frac{1}{T} = \frac{1}{T\_SIG}$$

Hence,

$$B\_P\_S = BAUD\_RATE$$

That is, the BAUD_RATE of the modulated carrier is equal to the number of bits per second entering the modem from the DTE. This, however, is only true for the simple modulation schemes we have discussed so far!

In all of the modulation techniques we have described, the controlled parameter of the carrier was only allowed to assume two different values. Suppose there were more than two possibilities for the value of the controlled parameter. Figure 9–7 illustrates the operation of an AM modem whose output can assume four different amplitude values (A1–A4). These four possibilities al-

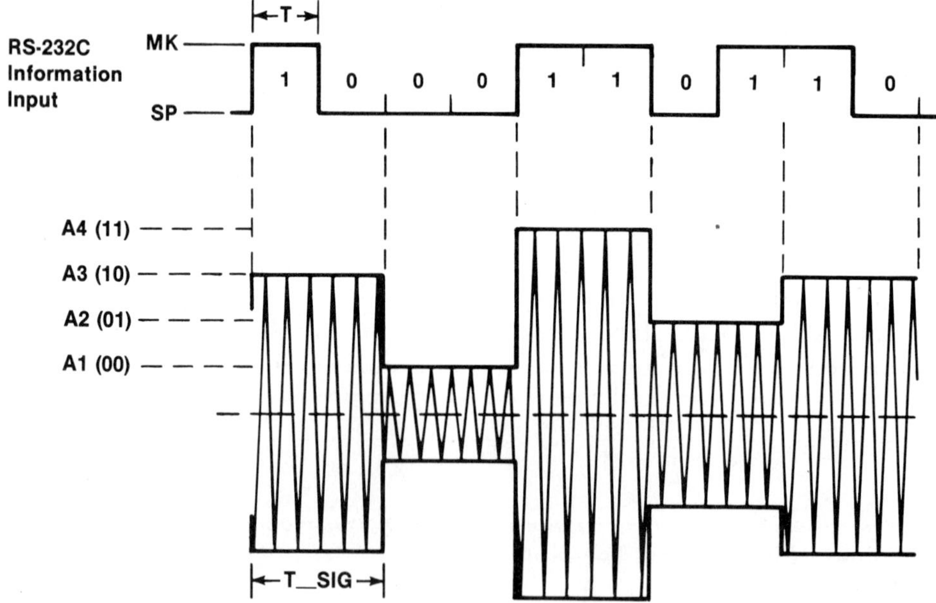

FIGURE 9–7   Multilevel Modulation

low us to represent a combination of two bits with each amplitude value. The assignment of bit patterns to amplitude values (which is completely arbitrary) is shown at the left side of Figure 9–7.

We take pairs of bits from the information input (these pairs are called DIBITS) and adjust the carrier's amplitude in accordance with the bit values. Notice that our minimum SIGNAL time is now twice the length of the bit interval on the RS-232C input.

$$2 * T = T\_SIG$$

$$\frac{1}{2 * T} = \frac{1}{T\_SIG}$$

$$\frac{1}{T} = 2 * \frac{1}{T\_SIG}$$

$$B\_P\_S = 2 * BAUD\_RATE$$

Each SIGNAL thus carries two information bits. The number of bits per second into (and out of) the modem is now two times the baud rate of the modulated carrier. We can now generalize this principle. If we define N as the *number of bits carried by each signal,* then

$$B\_P\_S = N * BAUD\_RATE$$

This equation says that if BAUD_RATE represents the number of SIGNALS being transmitted per second, and N represents the number of bits per SIGNAL, then the amount of data being moved by the carrier is

$$(bits\ per\ second) = (bits\ per\ signal) * (signals\ per\ second)$$

How can we relate N (the number of bits carried by each signal) to the number of possible SIGNALS we can produce by selecting different values for our modulation parameter? If we define P as the number of possible SIGNALS our modulator can produce, we have

$$P = 2^N$$

This is similar to the way we determined the number of possible characters in a data code with N data bits (see Chapter 5). When we solve the equation for N, we get a base 2 logarithm:

$$N = \log_2 P$$

(If you recall the definition of logarithms, this equation states that N is the exponent to which 2 (the base) must be raised to produce P.) We can now substitute for N in the equation for B_P_S:

$$B\_P\_S = (\log_2 P) * BAUD\_RATE$$

The following brief table of base 2 logarithms is handy:

NUMBER OF POSSIBLE SIGNAL VALUES (P)	BITS PER SIGNAL N (= $\mathrm{LOG_2\ P}$)
2	1
4	2
8	3
16	4

Modems are never designed with a value of P (11, for example) that does not make N an exact integer. There are also no commercially available telephone modems where P is greater than 16, so the previous table is complete. On the RS-232C side of the modem there are only two possible signal states (MARK and SPACE). It is therefore possible to freely interchange the terms baud and bits per second at the RS-232C interface. This is what leads to most of the confusion about these terms.

Let us do a sample calculation.

A DPSK synchronous modem can produce four different phase shift values ($+45°$, $+135°$, $+225°$ and $+315°$). If the RS-232C input is 2400 bits/s, what is the baud rate of the modulated carrier?

For four possible phase shifts, P = 4 and, from our table, N = 2. Then

$$B\_P\_S = N * BAUD\_RATE$$
$$2400 = 2 * BAUD\_RATE$$

and,

$$BAUD\_RATE = 1200$$

As a result of what we have learned in this chapter, we can very concisely summarize the operation of most commercially available modems:

TYPE	SPEED RANGE (B_P_S)	MODULATION TECHNIQUE	BITS PER SIGNAL (N)
Asynchronous	0–1800	FSK	1
Synchronous	1200–9600	DPSK	2, 3, or 4

(Actually, 9600-bit/s synchronous modems use a combination of DPSK and amplitude modulation. Rather than sixteen different phase shifts, these modems use twelve different phase shifts, four of which come in two different sizes. The reason for this is the difficulty in distinguishing between closely spaced phase shift values.)

9·4 MODULATION AND BANDWIDTH

We have previously stated that modulation was necessary to make digital signals fit into the 300-Hz–3300-Hz bandwidth of a "voice-grade" telephone circuit. We will now briefly delve into the subject of analog modulation to gain an understanding of how and why this technique works.

Figure 9–8A illustrates a carrier (frequency F_C) modulated by another sine wave (frequency F_M). Any waveform, this one included, has a frequency spectrum. The spectrum is a plot of the power in the waveform as a function of frequency. In the case of our AM signal, the spectrum consists of three distinct frequencies:

- the carrier F_C
- a sum frequency $F_C + F_M$
- a difference frequency $F_C - F_M$

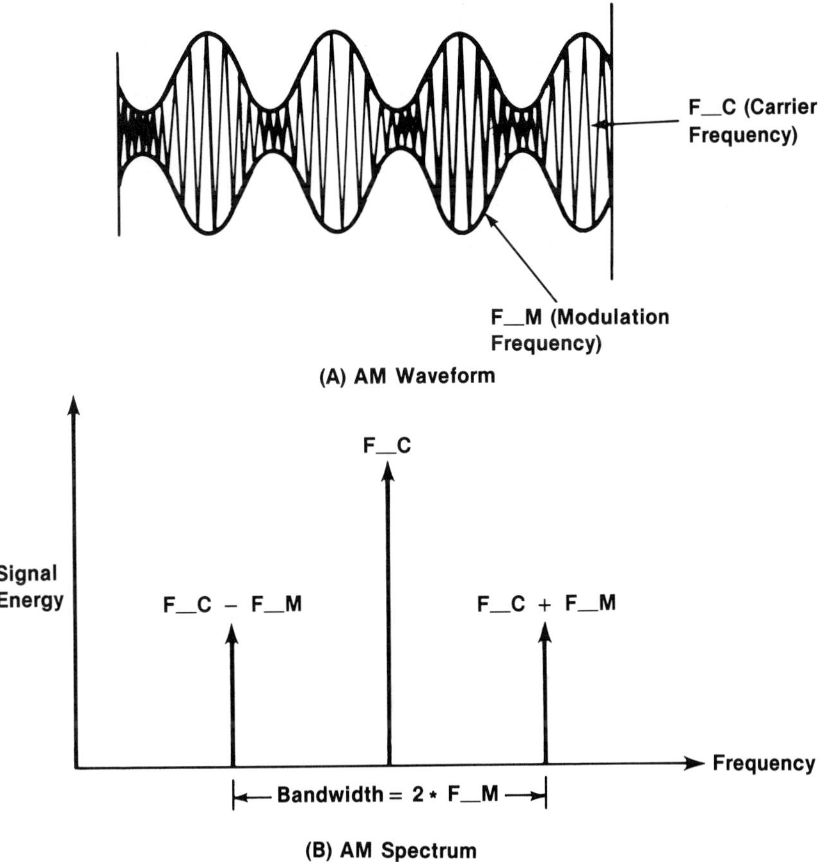

FIGURE 9–8 Analog Modulation

The sum and difference frequencies are called *sidebands*. For example, the spectrum of a 1000-Hz carrier modulated by a 200-Hz sine wave contains energy at 800 Hz, 1000 Hz, and 1200 Hz.

We state this without mathematical proof. If you are curious, a simple proof involving a few trigonometric identities is given in many electronics texts. There are two important conclusions we can draw from the spectrum:

- The modulating frequency (F_M) does not appear anywhere in the spectrum of the modulated carrier.
- The bandwidth of the modulated carrier (CAR_BW), which is defined as the difference between the highest and the lowest frequencies present, is

$$CAR\_BW = (F\_C + F\_M) - (F\_C - F\_M) = 2 * F\_M$$

From our last example we have F_M = 200 and

$$CAR\_BW = 1200 - 800 = 400 = 2 * F\_M$$

From this equation we can see that the carrier's bandwidth (that is, the width of its spectrum) is directly proportional to F_M, the rate at which the carrier is modulated. We could go through similar presentations for FM, PM, or any other modulation technique and come to similar conclusions. We can make the following general statement (again without proof).

For any modulation technique:

- The modulating frequency appears in the spectrum of the modulated carrier as two sidebands.[1] The separation of the sidebands from the carrier depends on the modulation technique being used.
- As the rate of modulation increases the total bandwidth occupied by the modulated carrier also increases.

All of this information applies to carriers modulated by sinewaves. What about our digital signals? Once again we state a mathematical result without proof:

It can be shown that any waveform shape is composed of the sum of a number of sine waves plus a DC component. (DC is considered a sine wave with frequency equal to zero.)

Suppose we have a TTL waveform consisting of alternating 1s and 0s at a rate of 100 bits/s. Figure 9–9 shows this waveform and its spectrum. The spectrum consists of the DC component and a series of sine waves at distinct frequencies. These frequency values are related to the bit rate of the digital waveform. For the alternating 1/0 pattern we have chosen, there will be a *fundamental frequency* (50 Hz) at one-half the basic bit rate and *harmonic frequen-*

---

1. Some modulation techniques produce only one sideband. These techniques are not typically used in telephone modems.

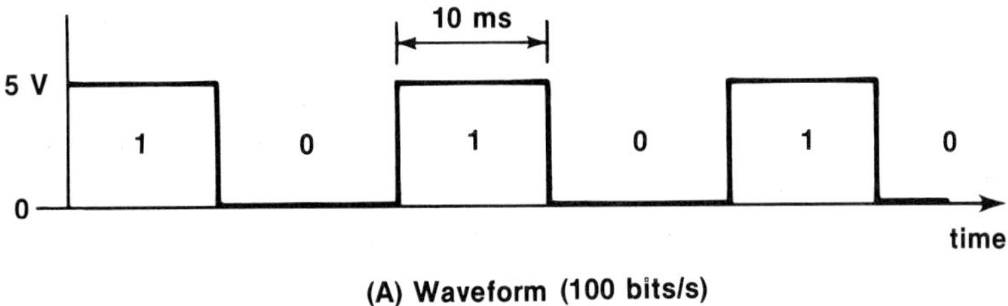

(A) Waveform (100 bits/s)

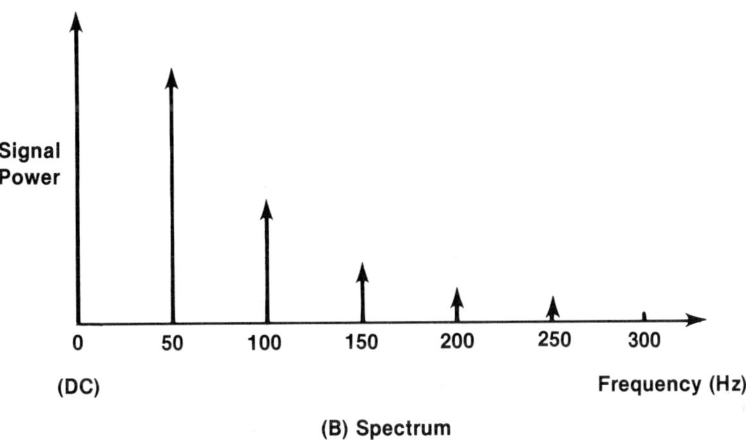

(B) Spectrum

FIGURE 9–9 The Spectrum of a Digital Waveform

cies at integer multiples of the fundamental. Notice that the power in the harmonics drops rapidly as frequency increases. If we had chosen to look at the spectrum of a similar signal on a RS-232C circuit, the DC component would be missing because the signal would spend equal amounts of time at voltages above and below ground. Otherwise the spectrum would be identical.

Alternating 1/0 patterns are not particularly useful for transmitting information. Actual digital signals are much more random, but for our 100-bit/s example, the 10-ms bit interval will still hold true. The spectrum of a more realistic 100-bit/s digital waveform would not be terribly different from what is illustrated in Figure 9–9B. Each of the discrete frequency components would be "smeared" toward the left into a more or less continuous band of frequencies. The overall trend of diminishing power with increasing frequency will remain the same. We would also get DC components on RS-232C circuits because the signal may contain a higher percentage of MARK than SPACE (or vice versa).

We can therefore conclude that almost all of the power in our 100-bit/s signal lies between DC and 200 Hz. There is theoretically an infinite number of harmonics, but those at higher frequencies are too small to measure, and, as a rule of thumb, we can say that the bandwidth of a digital signal is approximately two times its bit rate:

$$BASE\_BW = 2 * B\_P\_S$$

We have used the symbol BASE_BW because these digital waveforms are called *baseband signals* in literature about modulation.

What would actually happen if we tried to send a 100-bit/s digital signal down a telephone line? We have already concluded that all of its energy lies below 200 Hz. Since a phone line will not transmit anything below 300 Hz, the signal will get nowhere. Suppose we amplitude modulate a 2300-Hz carrier with our 100-bit/s signal. The resulting spectrum is illustrated in Figure 9–10. Each component of the baseband signal turns into a pair of sidebands around the 2300-Hz carrier. We have neatly translated our baseband spectrum up to a range that will easily pass through a telephone circuit. Here, in a few words, is the essence of this entire long-winded discussion:

> *Modulation* is a process which translates the spectrum of a baseband signal to a frequency range that can be transmitted on a given communications medium.

Suppose we increase the data rate of the baseband signal to 500 bits/s. We now have BASE_BW = 1000 Hz. If we continue to use the 2300-Hz carrier, the upper end of the spectrum of the modulated carrier (F_C + BASE_BW) is 3300 Hz, the upper limit of the phone line's frequency range. Can we continue to increase the data rate? Even if the carrier is moved to the exact center of the

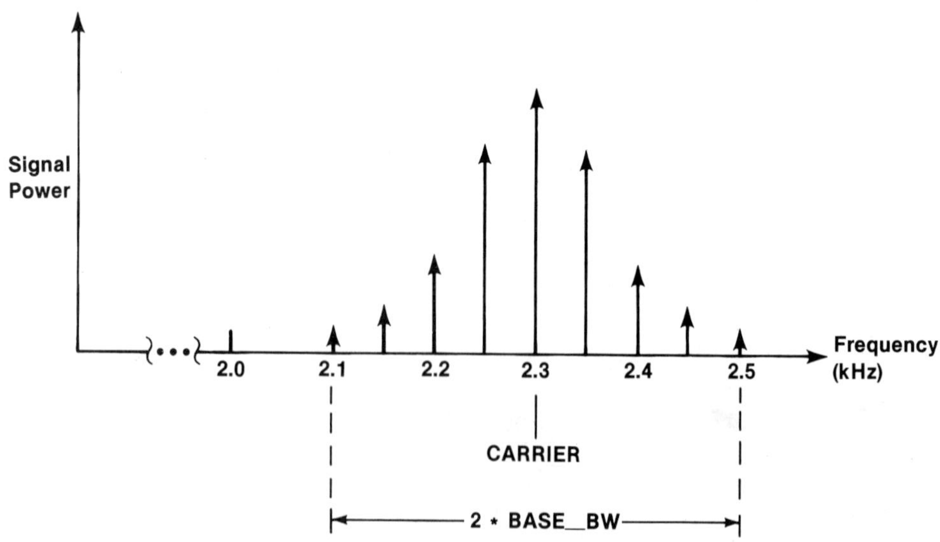

FIGURE 9–10

telephone band, with the simple AM scheme we have been using, a data rate of 750 bits/s will occupy the entire channel. How is it possible to design modems that send 9600 bits/s on a telephone circuit? The answer is multilevel modulation techniques. A 16-level modulation scheme, for example, would produce a baud rate that was lower than B_P_S by a factor of four. Since the baud rate (the number of signals transmitted per second) is exactly the rate at which the carrier is being modulated (recall Figure 9–7), the bandwidth is determined by BAUD_RATE rather than B_P_S. Stated another way, multilevel modulation techniques allow data rates to be substantially increased while maintaining the bandwidth of the modulated carrier within the confines of the telephone circuit.

Figure 9–11 illustrates the way several types of commercially available modems allocate the bandwidth of a telephone circuit.

FULL-DUPLEX, 4-WIRE Each directional pair of the 4-wire circuit has the full 3000-Hz bandwidth. The modems can therefore simultaneously modulate two carriers at full speed. The carrier frequencies are the same for both directions. Typical devices are synchronous and operate at speeds of 2400–9600 bits/s.

FULL-DUPLEX, 2-WIRE 3000 Hz of bandwidth is available on the 2-wire circuit, and it must be shared equally by the two directions. Each modem must transmit a different carrier frequency. Since this type of operation is characteristic of switched circuits, the carriers are usually referred to as *originate* and *answer*. Typical devices are asynchronous. Older ones operate at 0–300 bits/s. The newer 212-type devices have a 1200-bit/s speed available in addition to the 0–300-bit/s range.

The Full-duplex VA212 Auto Dial Modem Can Transmit Asynchronous Data at 1200 bits/second.
Courtesy of Racal-Vadic

HALF-DUPLEX, 2-WIRE   These are the same devices that operated full duplex on 4-wire circuits. Since their modulated carriers occupy the entire telephone bandwidth, they must be configured for half-duplex operation on 2-wire circuits.

HALF-DUPLEX, 2-WIRE WITH REVERSE CHANNEL   In this case the modulated primary channel carrier occupies almost the entire bandwidth of the phone line. A small slice of bandwidth is reserved for the much slower reverse channel. Typical devices are asynchronous with the primary channel operating at 1200 bits/s and the secondary operating at 5 bits/s.

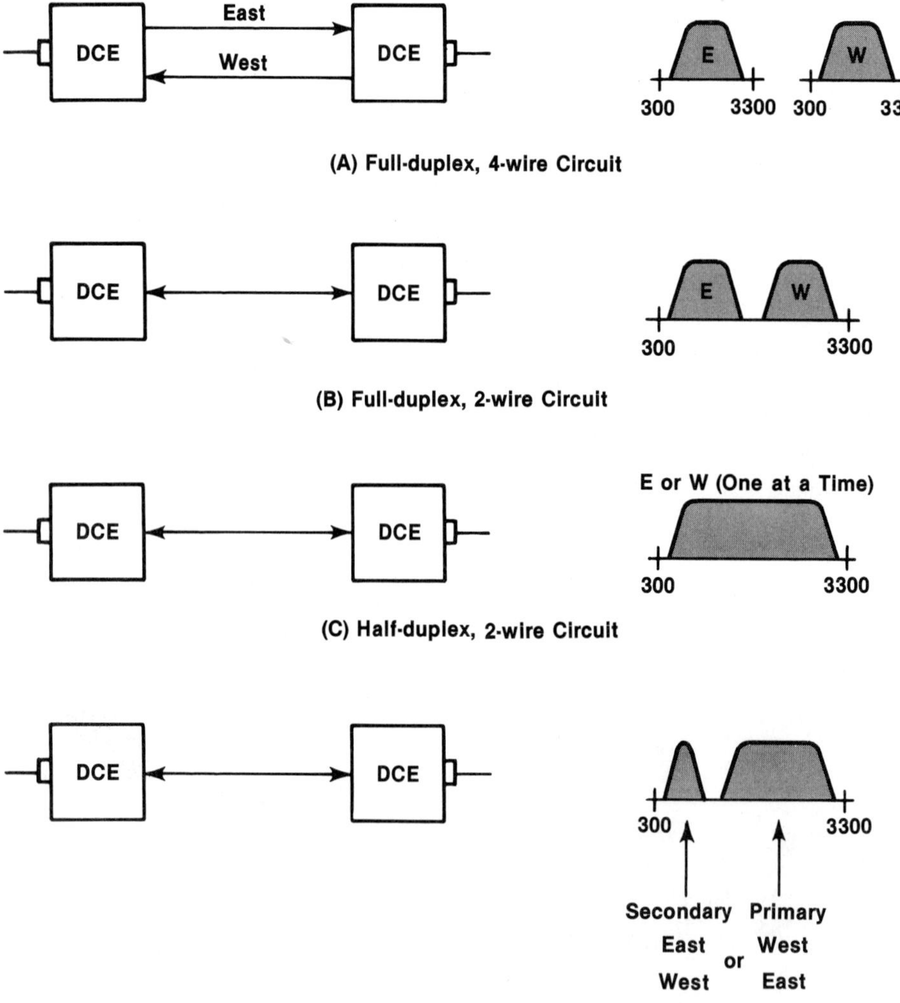

FIGURE 9–11   Bandwidth Allocation by Modem Type

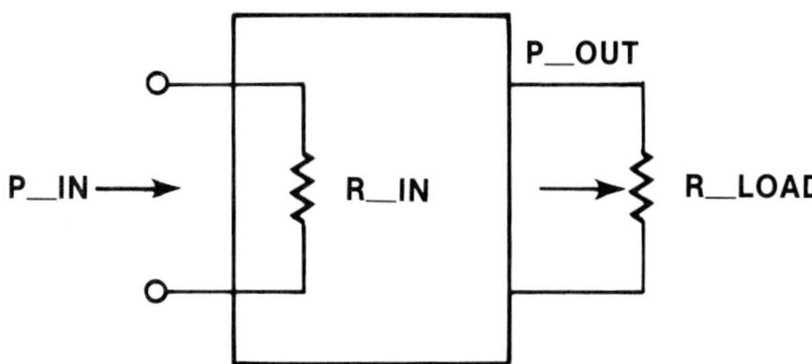

FIGURE 9–12   A Generalized Signal Transmission Device

9·5   SIGNAL STRENGTH

Until now we have said little about the strength of the carrier signal. As the carrier moves from one end of the telephone connection to the other, it encounters many different devices: modulators, amplifiers, hybrids, echo suppressors, microwave links, long lengths of copper wire. Some of these will strengthen the signal and some will weaken it. We can consider all of these devices "black boxes" with a block diagram such as the one illustrated in Figure 9–12. We are not now concerned with what happens inside this box. We simply measure the power entering it (P_IN) and the power leaving it (P_OUT). The *gain* of the box is defined as

$$\text{GAIN} = \frac{\text{P\_OUT}}{\text{P\_IN}}$$

We usually think of gain in relation to an amplifier, where the signal is strengthened and P_OUT/P_IN is greater than 1. In telephone circuits we will also encounter devices that weaken (*attenuate*) signals. We still refer to this as a gain. Its value, however, will be less than 1.

We must also be careful about how the power measurement is defined. Power may be measured at a specific frequency or as the sum of the power in all frequency components that are present. If the former method is used, the gain may vary significantly over the bandwidth of the phone line. In most of our discussions we will be talking about total signal power. In any event, measurement techniques must be consistent from input to output.

Suppose we connected two unspecified devices, output to input, as illustrated in Figure 9–13. Each device has a measurable gain. If we apply a signal to

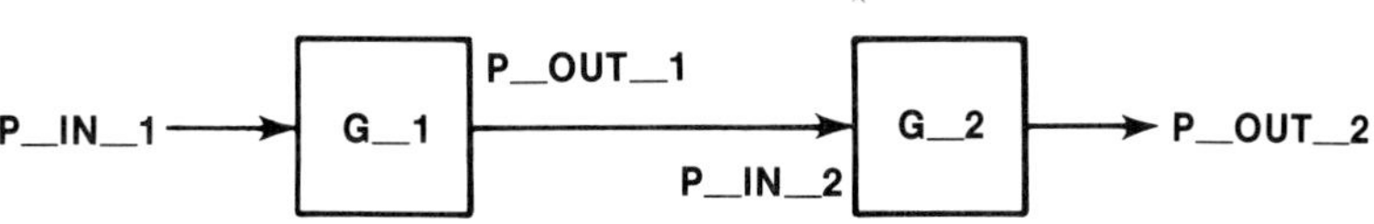

FIGURE 9–13   Cascaded Transmission Devices

the first input and compute the power at the last output, we can then divide the numbers to find the gain of the cascaded pair of devices.

$$P\_OUT\_1 = G\_1 * P\_IN\_1$$
$$P\_OUT\_2 = G\_2 * P\_IN\_2$$

Since

$$P\_OUT\_1 = P\_IN\_2$$

we can substitute the first equation into the second:

$$P\_OUT\_2 = G\_2 * (G\_1 * P\_IN\_1)$$

Therefore,

$$GAIN = \frac{P\_OUT\_2}{P\_IN\_1} = G\_1 * G\_2$$

We can easily generalize this to any number of cascaded devices with individual gains G_1, G_2, . . . , G_N:

$$G\_TOTAL = G\_1 * G\_2 * G\_3 * \cdot \cdot \cdot * G\_N$$

For example, if we had an amplifier with a power gain of 100 driving a filtering device which passed 90% of its input power to its output, the total gain is then:

$$\begin{aligned}
\text{(amplifier)} \qquad G_1 &= 100 \\
\text{(filter)} \qquad G_2 &= 0.9 \\
G_TOTAL &= (100) * (0.9) = 90
\end{aligned}$$

In the literature about telephone circuitry, gain is usually expressed not as a ratio but in a logarithmic unit called the *decibel* (dB). Its definition is

$$GAIN\_dB = 10 \log \left(\frac{P\_OUT}{P\_IN} \right)$$

This is a base ten logarithm. (Whenever the base of a logarithm is not explicitly given, assume that the base is ten.) According to the properties of logarithms, GAIN_dB is positive for amplifiers and negative for attenuators (and obviously 0 for devices that cause no change in signal level). For our amplifier with a power gain of 100,

$$GAIN\_dB = 10 \log(100) = 20 \text{ dB}$$

Since the log of a product is the sum of the logs of the individual terms in the product, the decibel gain for cascaded devices may be computed by addition of the gains of the components:

$$G\_TOTAL\_dB = G\_1\_dB + G\_2\_dB + \cdots + G\_N\_dB$$

We can also use decibels to measure absolute power levels as well as ratios. In order to do this we must define a *reference power level*. The most commonly used reference is 1 mW. The standard unit is *decibels relative to 1 mW* (dBm):

$$dBm = 10 \log\left(\frac{\text{signal power (mW)}}{1\,mW}\right)$$

Now that we have a logarithmic measure for power, we can go back to our original equation for GAIN_dB and do the following:

$$GAIN\_dB = 10 \log\left(\frac{P\_OUT/1\,mW}{P\_IN/1\,mW}\right)$$

$$= 10\left[\log(P\_OUT/1\,mW) - \log(P\_IN/1\,mW)\right]$$

$$= P\_OUT\_dBm - P\_IN\_dBm$$

Now solve for P_OUT:

$$P\_OUT\_dBm = P\_IN\_dBm + GAIN\_dB$$

We can easily generalize this to a cascade of devices, giving us a simple way to compute the output power if we know the input power and the individual device gains:

$$P\_OUT\_dBm = P\_IN\_dBm + G\_1\_dB + G\_2\_dB + \cdots + G\_N\_dB$$

As an example, look at Figure 9–14. We have two modems separated by a length of cable, an amplifier, and a second length of cable. With the gain figures shown, what is the power level arriving at the second modem if the first modem's carrier (the input signal to the transmission system) is operating at 0 dBm?

$$P\_OUT\_dBm = 0\,dBm - 15\,dB + 20\,dB - 10\,dB = -5\,dBm$$

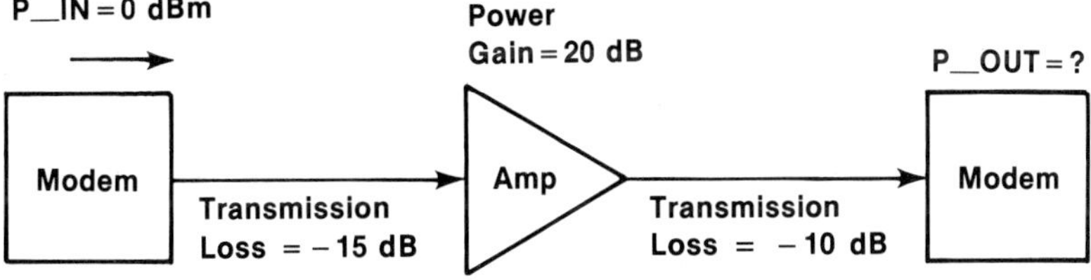

FIGURE 9–14 Example of End-to-End Transmission Calculations

We can compute a device's GAIN_dB from the output/input voltage ratio as well as from the power ratio. In this case the formula is

$$\text{GAIN\_dB} = 20 \log \left(\frac{\text{V\_OUT}}{\text{V\_IN}} \right)$$

As an example, what is GAIN_dB for a device whose output voltage is 10% of its input voltage?

$$\text{GAIN\_dB} = 20 \log(0.1) = -20 \text{ dB}$$

Notice that the constant multiplying the log is now 20 rather than 10 (as it was in the power formula). The reason for this is as follows:

When signal transmission elements are cascaded on a telephone line, all devices will frequently be designed with the same input resistance. All devices will therefore have R_IN = R_LOAD (see Figure 9–12). If we remember the relationship between power and voltage,

$$P = \frac{V^2}{R}$$

and the fact that $\log X^2 = 2 \log X$, a little algebra will demonstrate that if input and load resistances are equal, we get the same value for GAIN_dB from either a voltage measurement or a power measurement (see Review Question 9–11).

9·6 SIGNAL QUALITY AND LINE CONDITIONING

As a carrier is propagated over long distances and through the kinds of devices we have been discussing, its quality begins to deteriorate. The factors contributing to this deterioration are called *transmission impairments*. There are many varieties of transmission impairments. We will describe the four major types. These impairments cause modem receivers to incorrectly demodulate the incoming carrier, resulting in data errors. The effects of impairments become greater as we increase either the baud rate or the number of levels used by the modulation scheme. It is these impairments that determine the maximum data rate (bits/s) at which a particular modulation scheme is useful.

Noise

On telephone lines, as on any other electronic circuit, there is a constant, low-level background of electronic noise. This noise has a power spectrum distributed across the entire bandwidth of the phone line. The quantitative measure of this impairment is the ratio of signal power to noise power (commonly called *signal to noise ratio* or SNR):

$$SNR = \frac{SIGNAL\_POWER}{NOISE\_POWER}$$

It is frequently expressed in dB. Taking the log of both sides of the equation we get

$$SNR\_dB = SIGNAL\_dBm - NOISE\_dBm$$

As signals weaken over long transmission distances, they get closer to the background noise level and SNR decreases. The noise adds and subtracts random quantities from the instantaneous value of the carrier voltage and makes it more difficult for the demodulator to measure the carrier's modulated parameter. On standard telephone circuits in the United States, SNR is maintained at 24 dB or better.

Attenuation Distortion

We have repeatedly stated that telephone circuits can transmit frequencies between 300 Hz and 3300 Hz. The actual situation is not that simple. Because the end-to-end gain will vary over the bandwidth of the line, some frequencies are transmitted better than others. For standard telephone circuits the variation can be on the order of 8 dB in the center of the band (1000–3000 Hz) and even worse at the edges of the band. These effects limit the allowable spectrum of a modulated carrier.

Delay Distortion

In a manner similar to gain, the speed of signal propagation varies as a function of frequency over the telephone bandwidth. The result is a difference in transmission delay between frequencies in the center of the band and those at the edges. The difference can be as large as 2 ms on standard phone lines. While delay distortion can change the shape of a waveform because it shifts the relative phase of its frequency components, it does not introduce any new frequency components.

Harmonic Distortion

Harmonic distortion is caused by "nonlinearities" in devices such as amplifiers. These devices can change the shape of a waveform by effects such as "clipping" the peaks of the carrier wave. This type of distortion introduces frequency components (harmonics) that were not present in the carrier when it was initially transmitted. The new frequencies have effects similar to noise.

When we use switched circuits, the transmission characteristics will vary from one call to the next even if we repeatedly connect the same two subscriber loops. Since we have no control over which trunks will be selected, we cannot improve transmission characteristics beyond the specification of the "standard" circuit (known as a *type 3002 line* in the United States). For a dedicated circuit, however, we use the same facilities as long as we continue to lease the line. In this case the telephone company will, for a price, make adjustments to circuit components and improve transmission characteristics. This is known as *line conditioning.*

In the United States two varieties of conditioning are available for dedicated circuits:

1. C-type conditioning. C-type conditioning controls attenuation distortion and delay distortion. It is available in three levels (C1, C2, and C4), each of which places successively tighter tolerances on these two types of distortion.
2. D-type conditioning. D-type conditioning controls noise and harmonic distortion. It is available in only one level. SNR is improved by at least 4 dB, and the level of harmonics is reduced by at least 10 dB.

Line conditioning can frequently provide substantial increases in the data capacity of dedicated circuits. This was especially true for older modem designs. For example, the Bell 202, an asynchronous modem, is rated at 1200 bits/s on unconditioned lines and 1800 bits/s on lines with C2 conditioning. Newer designs have been able to achieve 9600 bits/s (synchronous) on switched circuits, but their error rates are fairly high.

9·7 A THEORETICAL UPPER LIMIT TO CHANNEL CAPACITY

We present another theoretical result without proof. This result was derived by Claude Shannon in 1948, and it is the fundamental theorem of a science called Information Theory. It states that there is an absolute upper limit to the rate at which data can be sent across any communications channel. This rate depends on the bandwidth of the channel (CHAN_BW) and the signal to noise ratio (SNR). The formula for this *channel capacity* (in bits per second) is

$$\text{CHAN\_CAP} = \text{CHAN\_BW} * \log_2(1 + \text{SNR})$$

Regardless of how clever the modulation scheme or how precisely designed the demodulator circuits, this limit cannot be exceeded.

We can get an intuitive idea of why this is true by recalling the equation in Section 9·3 that allowed us to compute B_P_S, the data flow produced by a given modulation technique:

$$B\_P\_S = BAUD\_RATE * \log_2 P$$

where P is the number of possible SIGNAL levels our modulated carrier can assume. There is a strong parallel between these two equations. CHAN_CAP actually represents the maximum value B_P_S can ever attain on a given medium. B_P_S depends on BAUD_RATE and P. Since an increase in BAUD_RATE widens the spectrum of the modulated carrier, CHAN_BW is obviously a limiting factor on BAUD_RATE. We have also stated that the primary effect of noise is to increase the difficulty of making precise measurements on the modulation parameter of a carrier. SNR is therefore a limiting factor on P since, as we increase the number of possible signal values (phases, for example), they become so closely spaced that noise will make it impossible to distinguish one value from another.

What is this limit for a typical switched telephone line? We have

$$CHAN\_BW = 3300 - 300 = 3000 \text{ Hz}$$

and

$$SNR\_dB = 10 \log SNR = 24 \text{ dB}$$

Hence,

$$SNR = 250$$

Therefore,

$$CHAN\_CAP = 3000 * \log_2(251) .$$

Since $\log_2 (256) = 8$, which is close enough to $\log_2 (251)$ for this calculation,

$$CHAN\_CAP = 3000 * 8 = 24\,000 \text{ bits/s}$$

In recent years some commercially available modems have been able to achieve 19 200 bits/s on dedicated voice-grade telephone circuits. These devices usually have the optional ability to operate at lower speed if error rates are intolerable. For switched circuits no commercially available modem has been able to operate in excess of 9600 bits/s.

9·8 PULSE CODE MODULATION

As a final note on this subject, we briefly describe a process called *pulse code modulation* (PCM). The operation of PCM depends on the fact that the shape of an analog waveform can be perfectly reconstructed from a set of samples of its amplitude. The only restriction is the requirement that the sampling rate must be at least 2 times the highest frequency in the spectrum of the analog signal. The concept and its associated waveforms are illustrated in Figure 9–15.

The analog signal (waveform A) is fed to the input of a *sampler* circuit which periodically takes a "snapshot" (waveform B) of the signal's amplitude and passes it to an *analog-to-digital (A/D) converter*. The A/D converter measures the voltage of each sample and produces a corresponding binary representation of that voltage. Each sample is converted into a fixed number of bits (the larger the number of bits per sample, the more accurate the representation of the analog waveform). These digitized samples are then transmitted serially (waveform C) using digital circuits and media capable of carrying these baseband signals.

The serial bit stream is received by a *digital-to-analog (D/A) converter* whose output is an analog voltage determined by the binary signals at its input. Since the D/A converter receives no information regarding the amplitude of the original analog signal at times other than the sample points, its output is a series of voltage steps (waveform D) which roughly follow the variations of the original signal. When the steps are passed through an electronic *filter* circuit, however, the result is an exact reproduction of the original signal (waveform E).

The analog waveform can be voice, modem output, or any other similar signal. In recent years many sections of the U. S. telephone network have been converted to PCM operation. This is particularly true of the equipment in switching offices. There are several reasons for this:

- Digital integrated circuits are inexpensive and readily available.
- Digital circuits have an inherent noise immunity. (Precise voltage measurements are not required, only a determination of whether the signal is 1 or 0.)
- The time interval between samples of one analog signal can be used to carry samples of other signals. This process is called *time division multiplexing* (TDM). We will have more to say about this subject in Chapter 12.

It is somewhat ironic to note that data communications requires the use of modems to convert baseband digital signals to a different form so they can be carried by a telephone network which, in many places, converts everything back to a digital format. In fact, it is sometimes possible to lease end-to-end digital facilities from the phone company, allowing data links to be constructed without modems. Nevertheless, the bandwidth of ordinary switched telephone connections remains 300 Hz to 3300 Hz. Until the telephone network becomes totally digital, modems will continue to be required.

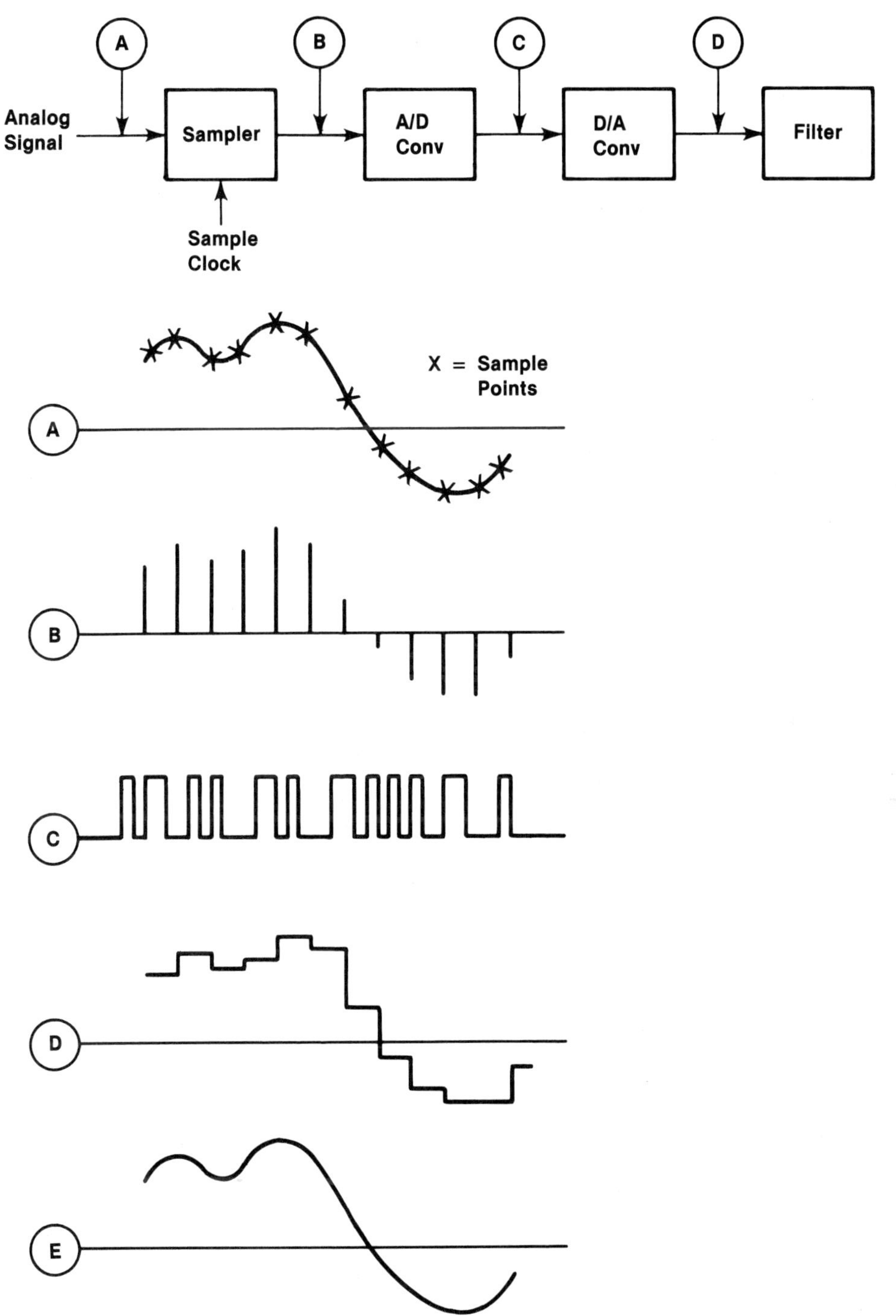

FIGURE 9–15 Pulse Code Modulation/Demodulation

REVIEW QUESTIONS

1. What is the essential difference between modulation techniques for the synchronous data format and for the asynchronous data format?

2. Could we define a differential frequency shift keying (DFSK) technique for synchronous data that modulated the F parameter in a manner analogous to the way DPSK modulates phase? What would happen in this experimental technique if we attempted to transmit a string of 1000 consecutive MARK bits?

3. Suppose a carrier was a train of flat-topped pulses rather than a sine wave. What parameters of the pulse train might be modulated to carry information?

4. An FSK modem is designed to transmit eight different frequency signals. How many bits are represented by each transmitted signal? If RS-232C data is entering this modem at 4800 bits/s, what is the baud rate of the modulated carrier?

5. A DPSK modem which can produce four different phase shifts is transmitting its modulated carrier at 600 baud. What is the data rate in bits per second?

6. Is it possible to build a 2-wire full-duplex telephone modem with each carrier using a bandwidth of 2000 Hz? Explain why.

7. What is the approximate width of the baseband spectrum of a 200-bit/s digital signal? If we use simple amplitude modulation to put this signal on a 2000-Hz carrier, what are the upper and lower frequency limits of the resulting spectrum?

8. A DPSK modem is designed as follows:

DIBIT	PHASE SHIFT
00	$+ 45°$
01	$+135°$
10	$-135°$
11	$-45°$

Assuming that we start at a phase vector of $0°$, what is the state of the phase vector after the sequence 11010110 is transmitted?

9. Express the following in dB:

POWER RATIO		VOLTAGE RATIO	
a.	0.001	d.	8.0
b.	0.040	e.	0.5
c.	500.000	f.	800 000.0

10. Express the following as a power ratio:

    a. -20 dB
    b. $+ 7$ dB
    c. -37 dB
    d. $+43$ dB
    e. $+16$ dB

11. An amplifier has a 1000-ohm input resistance and drives a 1000-ohm load. A 2-V signal is applied to its input and a 20-V signal is measured at its output. Compute GAIN_dB from the voltage ratio. What is the power at the input and at the output in milliwatts? Compute GAIN_dB from the power ratio.

12. Refer to Figure 7–6. West subscriber is a modem with a transmitted carrier level of -5 dBm and the circuit components have gains as follows:

West hybrid	-5 dB
Eastbound echo suppressor	-2 dB
Link from echo suppressor to eastbound amplifier	-8 dB
Eastbound amplifier	$+30$ dB
Link from eastbound AMP to East hybrid	-8 dB
East hybrid	-5 dB

What is the power level (dBm) at East subscriber? (Assume that all other links have no gain or attenuation.)

13. What is the capacity of a 300-Hz to 3300-Hz channel with SNR $=$ 7? What is the capacity if the SNR is increased to 15, to 31? What can you conclude about attempting to increase channel capacity by simply increasing the power level of the carrier?

chapter ten

ERROR DETECTION

In Chapter 2, when the format of the asynchronous character was introduced, we briefly mentioned parity and its function as an error detection device. In order to gain the ability to detect errors, we had to transmit more bits than were absolutely necessary to convey the information content of our message. This concept is called *redundancy*, and it is the basis for all communications error detection methods. For example, you can usually understand an English sentence in spite of a few misspelled words. This is due to the redundancy in the language. If English words were "encoded" by using only the minimum number of letters necessary to distinguish one word from another, error detection would be difficult because changing a single letter would always produce a word with a totally different meaning.

10·1   PARITY IS NOT GOOD ENOUGH

Parity is only the first, and the simplest, error detection scheme we will discuss. The more complex techniques were invented because parity is not a perfect error detection method. Consider the following 5-bit character transmitted with odd parity:

$$0 \quad 0 \quad 1 \quad 1 \quad 0 \quad 1$$
$$\uparrow$$
parity bit

The parity bit is inserted by the transmitter to maintain an odd number of 1s in the character. The receiver is also expecting to find an odd number of 1s in all characters and, if the count is even, will declare an error. Suppose one bit in the character is changed during transmission:

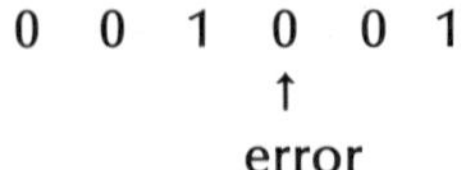

The count is now even and the error is detected. What happens if two bits are corrupted in the same character?

$$0 \quad 1 \quad 1 \quad 0 \quad 0 \quad 1$$
$$\uparrow \qquad \uparrow$$
$$\text{errors}$$

(Remember that an error can be a change in either direction.) This double error has returned the count of 1s to an odd number and the error goes undetected. This would be true regardless of whether we had chosen an odd or even parity scheme. For either choice, an even number of errors in the same character is undetectable. This is especially troublesome because errors on telephone lines tend to come in bursts.

10·2 THE BLOCK CHECK CHARACTER AND LRCS

These error bursts are a more serious problem at high data rates, and a number of techniques have been developed to improve the situation. At these speeds (typically 2400 bits/s and higher), telephone data transmission is usually synchronous and blocked. A block is a group of characters transmitted with no time gaps between them (this is the definition of synchronous data). Error detection techniques for blocks are based on the transmitter attaching a *block check character* (BCC) to the end of each message (see Figure 10–1).

The text block is a contiguous string of ASCII or EBCDIC characters. Block sizes vary, but they are typically in the area of 256 characters. At the end of the block there is an *end of message indicator* (EOM). EOM varies from one device to another, but, in general, it will be one or two control characters from the device's usual character set. The EOM indicator is a marker which informs the receiving device of the location of the block check character, which may be larger than normal text characters. The general procedure operates as follows:

- The transmitter uses a predefined algorithm to compute the BCC from the hexadecimal values of the message text characters.
- The resulting BCC is appended to the message following the EOM indicator.

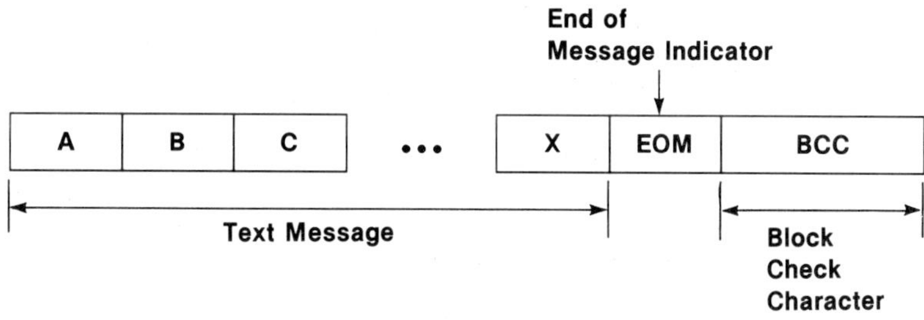

FIGURE 10–1 Blocked Message with BCC

- The receiver uses the same algorithm to compute its own value of the BCC from incoming data.
- If there are no errors, transmitted and received data are the same, and the result of the receiver's calculation should match the BCC attached to the end of the message block.
- If the results do not match, the message is assumed to contain errors.

Exactly what happens after the error is detected falls under the subject of protocols, which we will study in the next chapter. Protocols usually provide the receiver with a mechanism for requesting a retransmission of the erroneous message.

There are many different algorithms for producing a BCC. The most basic technique is the *longitudinal redundancy check* (LRC), which, despite its imposing name, is nothing more than an extention of parity into two dimensions. Consider the small message shown in Figure 10–2. The message is a block of 4-bit text characters, transmitted LSB first, and with each character having its own parity bit. In addition, each bit position of the BCC is an odd parity bit on the corresponding bit position in all text characters in the block. This is easy to see from the following tabulation of the message:

```
Odd Char
Parity
 ↓
0 0 1 0 0        Char 1
0 1 1 0 1        Char 2
1 0 0 1 1        Char 3
1 1 1 0 0        Char 4

1 1 0 0 1        BCC (odd)
```

In this presentation, each bit of the BCC is formed as an odd parity bit for the corresponding column. (The parity bit of the BCC is based on the BCC itself rather than the parity bits of the text characters.) Actual messages, of course, will be much longer than this illustration.

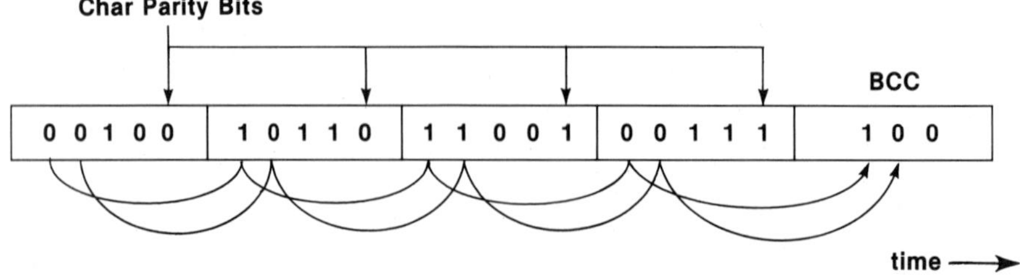

FIGURE 10–2 LRC Computation

Suppose we had a double bit error in char 2 (the errors are indicated by
< >).

```
0   0   1     0     0     Char 1
0   1   <0>   <1>   1     Char 2
1   0   0     1     1     Char 3
1   1   1     0     0     Char 4
─────────────────────────
1   1   0     0     1     BCC (Odd)
        ↑     ↑
          errors
```

As we noted previously, if we only had the character parity bits, this error
would be undetected. With the BCC it is very clearly indicated. If there was
only a single bit error, it could actually be corrected by finding the intersection
of the row and column containing the errors:

```
      0   0   1     0   0     Char 1
  →   0   1   <0>   0   1     Char 2
      1   0   0     1   1     Char 3
      1   1   1     0   0     Char 4
    ───────────────────────
      1   1   0     0   1     BCC (Odd)
              ↑
```

In practice this is not usually done. The receiver monitors for character
parity and LRC errors and if either or both occur, a retransmission is requested.
It is also possible that an error will occur in the BCC itself. These "false alarms"
are an unavoidable consequence of any error detection scheme, but their only
effect is an occasional unnecessary retransmission.

10·3   CYCLIC REDUNDANCY CHECK

A newer and more sophisticated family of algorithms for BCC calculation is
the *cyclic redundancy check* (CRC). It is by far the most effective error detection technique in common use. In this procedure the entire message block is
treated as one long binary number. It is "divided" by a binary constant and the
remainder is transmitted as the BCC. "Divided" is in quotes because the actual
binary division procedure is quite different from ordinary long division. Before
getting into that, we will illustrate the process with a decimal example. Suppose
our message was the sequence 893725. We select 99 as our divisor and arrive at
a remainder of 52. What actually gets transmitted is

89372552

Suppose lightning strikes the microwave tower and we receive a message
with two changed digits:

89462552

The receiver is programmed to recognize the last two digits as the BCC and everything preceding it as the message text (for clarity, EOM has not been shown). The receiver divides the number it believes to be the message (894625) by the agreed upon constant (99) and gets a remainder of 61, which is obviously different from what was transmitted.

Although the details are different, the steps of the binary division procedure parallel those of decimal division. We will therefore review the components of long division:

$$\text{DIVISOR}\ \overline{)\ \text{DIVIDEND}}^{\ \underline{\text{QUOTIENT}}}$$

$$\vdots$$

$$\overline{}$$

REMAINDER

The mathematical basis of the CRC procedure is very complex and beyond the scope of this book. Binary long division will therefore be presented as a "cookbook" procedure. From the details of the procedure we will get an insight into the reasons why the CRC is such an effective error detector. In our procedure the dividend will be based on the text of the message; the divisor will be the constant agreed upon by transmitter and receiver; we will arrive at a quotient which will be discarded and a remainder which will be the BCC.

First we must decide how many bits we will have in the BCC (call this BCC_SIZE). The rules of this game state that:

- The number of bits in the divisor must be BCC_SIZE + 1.
- The least significant bit of the divisor must be 1.

Suppose we want BCC_SIZE = 3. We must therefore select a 4-bit divisor. For our example we choose 1011.

The dividend must be created from the message text. More rules:

- List the message bits, in order of transmission, with the first bit of the message at left.
- At the right end of the message, add as many 0s as there are bits in the BCC (BCC_SIZE).

Suppose our message is:

$$1\ 0\ 1\ 0\ 0\ 1$$
$$\uparrow$$
first bit transmitted

The dividend is therefore

$$1\ 0\ 1\ 0\ 0\ 1\ 0\ 0\ 0$$

Our binary division example appears as follows:

$$1\ 0\ 1\ 1\ \overline{)1\ 0\ 1\ 0\ 0\ 1\ 0\ 0\ 0}$$

Before we go any further we must review the Boolean EXCLUSIVE OR (XOR) function because it is involved in the division process. In truth table form:

A	B	A XOR B
0	0	0
0	1	1
1	0	1
1	1	0

The result of XOR is 1 if the input bits are different and 0 if the inputs are the same. We now describe the steps taken by the transmitter to perform the division. Each step will be illustrated with the example message.

1. Starting from the left, scan the dividend until you find the first 1. Write the divisor below the dividend with the most significant bit of the divisor aligned with this 1.

$$1\ 0\ 1\ 1\ \overline{)\ 1\ \ 0\ \ 1\ \ 0\ \ 0\ \ 1\ \ 0\ \ 0\ \ 0}$$
$$\underline{1\ \ 0\ \ 1\ \ 1}$$

2. Place a 1 in the quotient directly above the least significant bit of the divisor. Then take the XOR of the divisor and the dividend bits directly above. Write the result of the XOR under the divisor.

$$1$$
$$1\ 0\ 1\ 1\ \overline{)\ 1\ \ 0\ \ 1\ \ 0\ \ 0\ \ 1\ \ 0\ \ 0\ \ 0}$$
$$\underline{1\ \ 0\ \ 1\ \ 1}$$
$$0\ \ 0\ \ 0\ \ 1$$

3. Write all remaining bits of the dividend in line with the results of the XOR. This gives a new dividend.

$$1$$
$$1\ 0\ 1\ 1\ \overline{)\ 1\ \ 0\ \ 1\ \ 0\ \ 0\ \ 1\ \ 0\ \ 0\ \ 0}$$
$$\underline{1\ \ 0\ \ 1\ \ 1}$$
$$0\ \ 0\ \ 0\ \ 1\ \ 0\ \ 1\ \ 0\ \ 0\ \ 0$$

4. Repeat Steps 1–3 with the new dividend.

$$11$$
$$1\ 0\ 1\ 1\ \overline{)\ 1\ \ 0\ \ 1\ \ 0\ \ 0\ \ 1\ \ 0\ \ 0\ \ 0}$$
$$\underline{1\ \ 0\ \ 1\ \ 1}$$
$$0\ \ 0\ \ 0\ \ 1\ \ 0\ \ 1\ \ 0\ \ 0\ \ 0$$
$$\underline{1\ \ 0\ \ 1\ \ 1}$$
$$0\ \ 0\ \ 0\ \ 1\ \ 0\ \ 0$$

5. When Step 1 cannot be executed without the divisor extending beyond the right end of the latest dividend (as is the case shown), the division is complete. The remainder is a BCC_SIZE group of bits selected from the last line of the division process. In our example the remainder is 100. This is the BCC and its *most significant bit is transmitted first.*
6. To complete the quotient, fill all empty quotient bit positions with 0 (except for BCC_SIZE positions at the left end of the quotient).

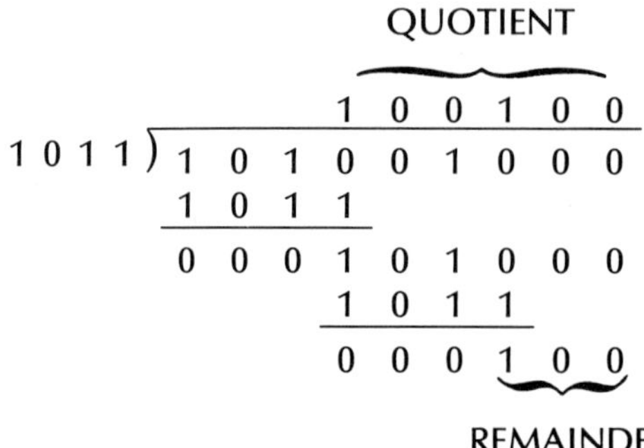

The remainder is now transmitted. The receiver performs this computation using the same divisor and, if there are no errors, gets the same remainder.

10·4   HARDWARE FOR THE CRC

Now that we can compute the CRC on paper, the next step is to design hardware to produce the same result. A block diagram for the transmit side of such hardware is illustrated in Figure 10-3. We have a Transmit Shift Register

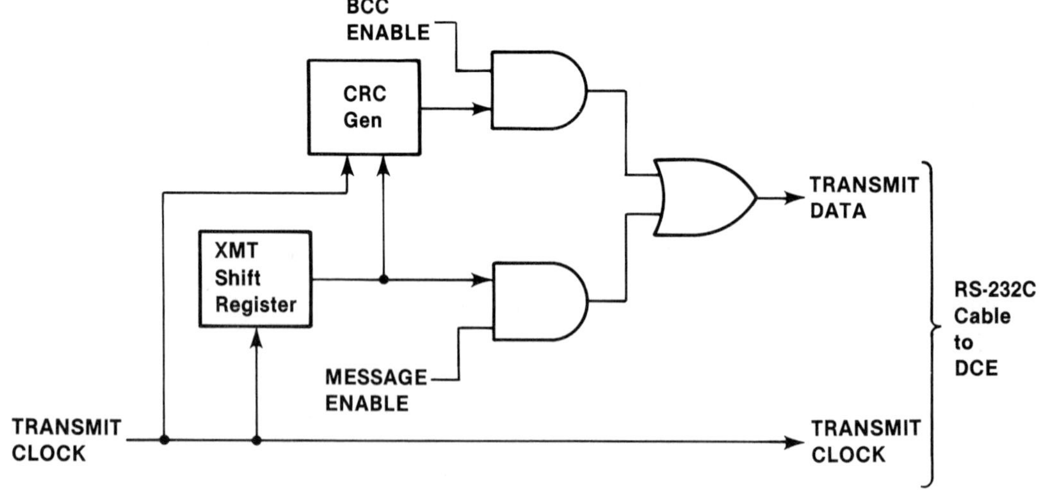

FIGURE 10–3   CRC Generation in a Transmitter

which is loaded with message text characters by parallel transfers from a micro-processor (similar to the UART transmit function). This is synchronous data, and the clock that shifts this data to the modem is also the RS-232C TRANSMIT CLOCK signal. While this parallel-to-serial conversion is happening, the MES-SAGE ENABLE signal is asserted (allowing the shift register output to be trans-mitted to the modem), and BCC ENABLE is negated (blocking the output of the CRC *generator*).

During the message transmission phase, the shift register output and the TRANSMIT CLOCK are also applied to the CRC generator. After the last mes-sage bit is shifted out to the modem, the CRC generator will contain the the ex-act remainder that would be produced by the long-division process described earlier. MESSAGE ENABLE will now be negated and BCC ENABLE asserted. The next BCC_SIZE clock pulses will produce the remainder at the output of the CRC generator and, therefore, on the TRANSMIT DATA line to the modem.

What is inside the CRC generator? For a rather complex procedure the hardware is remarkably simple. It is merely a collection of shift register stages and XOR gates. A general picture of a CRC generator is shown in Figure 10–4. The small squares are flip-flop shift register stages. In response to the TRANS-MIT CLOCK (which is not shown) the data contained in them shifts from left to right as usual. At various places a two-input XOR gate (symbolized by $\oplus$) sep-arates a pair of shift register stages. The last shift register stage enters an XOR gate whose other input is the serial data from the transmit shift register shown in Figure 10–3. The output of this last XOR gate is the *feedback* signal. It loops back to become the input of the first shift register stage and is also the second input to all the intermediate XOR gates.

CRC generators come in many different configurations. The number of shift register stages and the location of the XOR gates are determined by a mathematical entity called the *generator polynomial*. The generator polynom-ial, in turn, is derived from the divisor that was used in the long-division pro-cess. The choice of a divisor therefore uniquely determines the circuit diagram of the CRC generator.

A generator polynomial, like any other algebraic polynomial, can be rep-resented as a sum of terms. Each term is the product of a constant (called the

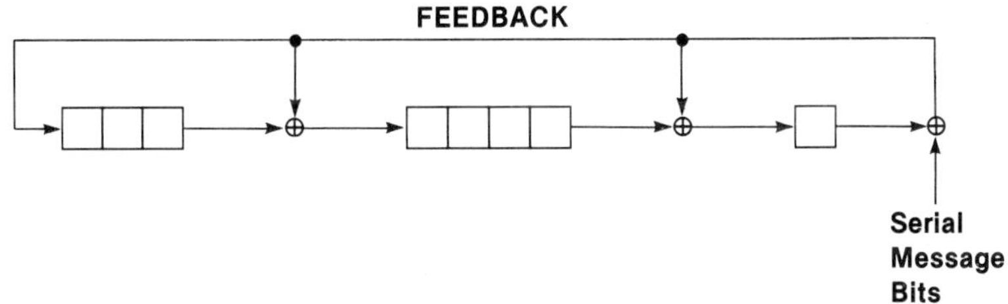

FIGURE 10–4 Generalized CRC Generator

coefficient) and a variable (we call our variable X) raised to an exponent. We assign an identifying number to each coefficient and symbolize the set of coefficients as follows:

$$C[0], C[1], C[2], \ldots, C[N-1], C[N]$$

A polynomial can then be represented as

$$C[N] * X^N + C[N-1] * X^{N-1} + \cdots + C[1] * X + C[0]$$

Since the highest exponent is N, we say that this polynomial is of degree N.

The only unusual thing about a generator polynomial is that its coefficients are binary. In fact, for any specific CRC generator, the coefficients of the generator polynomial are the bits of the divisor used in the equivalent long-division procedure. The least significant bit of the divisor is C[0], the next is C[1], and so on. For the divisor used in the example of the last section (1011), we have

$$C[0] = 1, C[1] = 1, C[2] = 0, C[3] = 1$$

The generator polynomial (degree 3) is

$$
\begin{aligned}
& C[3] * X^3 + C[2] * X^2 + C[1] * X + 1 \\
=\ & 1 * X^3 + 0 * X^2 + 1 * X + 1 \\
=\ & X^3 + X + 1
\end{aligned}
$$

Note that the degree of the generator polynomial is exactly equal to BCC_SIZE and, according to the rule defined in the previous section, C[0] is always 1.

How does this relate to flip-flops and XOR gates? We can inspect the generator polynomial and draw the circuit diagram according to the following rules (see Figure 10–5 for an example):

1. Start drawing the shift register stages from the left and number them, starting with 1. Each time the stage number is equal to an exponent in the generator polynomial, draw an XOR gate following that stage. (Ignore the 1 that always appears at the end of the polynomial and remember that X is really X^1.)
2. Serial message data is always the second input to the last (rightmost) XOR gate. The output of this gate is fed back to the input of stage 1.
3. The feedback signal also becomes the second input to all intermediate XOR gates.

To simplify the illustration, TRANSMIT CLOCK is not shown in Figure 10–5. It nevertheless appears as the clock input to all flip-flops.

Notice that the number of shift register stages is equal to the degree of the generator polynomial (which is also BCC_SIZE). After all message bits have been clocked through the CRC generator, each shift register stage will contain

$$X^8 + X^5 + X^3 + X^2 + 1$$

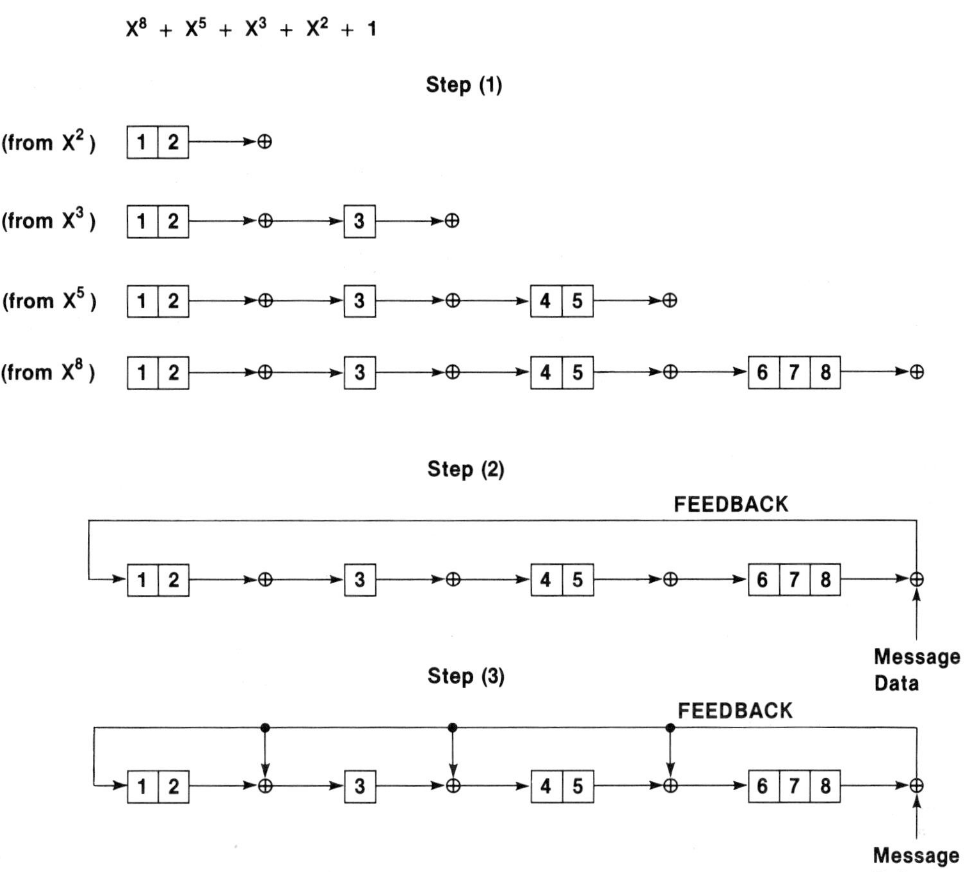

FIGURE 10–5 Circuit Diagram for a Generator Polynomial

one bit of the remainder. The least significant bit will appear in stage 1; the most significant bit in the stage at the right.

Let us take the division example of the previous section and follow it through its hardware implementation. The CRC generator for $X^3 + X + 1$ is illustrated in Figure 10–6, where we have labeled the stages S1, S2, and S3. We also use these symbols to represent the binary output value of the respective stages. Initially all stages will be cleared to 0. At any time we can determine the input to each stage from

$$S1\_IN = FB = S3 \ XOR \ MSG\_DATA$$
$$S2\_IN = S1 \ XOR \ FB$$
$$S3\_IN = S2$$

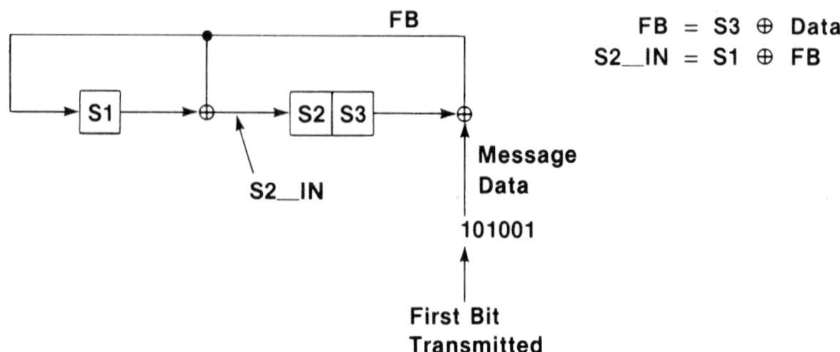

FIGURE 10–6   CRC Generator for $X^3 + X + 1$

Each time a bit appears at the message data (MSG_DATA) input, we apply a shift clock to all stages, and the binary value at each input becomes the new output of that stage. Since our message has six bits, after six clock pulses the remainder (BCC) will be available. The entire process is described in the following tables. The first table shows the initial condition of the shift register stages (before the first clock pulse is applied); later tables show their condition after each clock pulse. We also have columns for message data and the outputs of the two XOR gates (FB and S2_IN). Since we know the sequence of data bits, we can enter them in their column now.

CLOCK NUMBER	S1	S2_IN	S2	S3	MSG_DATA	FB
Start	0		0	0	1	
1					0	
2					1	
3					0	
4					0	
5					1	
6						

From this, we can compute

$$FB = S3 \text{ XOR } MSG\_DATA = 0 \text{ XOR } 1 = 1$$

and

$$S2\_IN = S1 \text{ XOR } FB = 0 \text{ XOR } 1 = 1$$

and enter these values in the table.

CLOCK NUMBER	S1	S2_IN	S2	S3	MSG_DATA	FB
Start	0	1	0	0	1	1
1					0	
2					1	
3					0	
4					0	
5					1	
6						

We are now ready to apply the first clock pluse. At this time, the shift register stages assume their new values:

$$S1\_IN = FB = 1 \rightarrow S1$$
$$S2\_IN = 1 \rightarrow S2$$
$$S3\_IN = S2 = 0 \rightarrow S3$$

These values are entered in the appropriate columns:

CLOCK NUMBER	S1	S2_IN	S2	S3	MSG_DATA	FB
Start	0	1	0	0	1	1
1	1		1	0	0	
2					1	
3					0	
4					0	
5					1	
6						

We now compute new values for FB and S2_IN:

$$FB = 0 \text{ XOR } 0 = 0$$
$$S2\_IN = 1 \text{ XOR } 0 = 1$$

and enter these in the table.

CLOCK NUMBER	S1	S2_IN	S2	S3	MSG_DATA	FB
Start	0	1	0	0	1	1
1	1	1	1	0	0	0
2					1	
3					0	
4					0	
5					1	
6						

After the second clock pulse, the new values for S1, S2, and S3 can be entered:

CLOCK NUMBER	S1	S2_IN	S2	S3	MSG_DATA	FB
Start	0	1	0	0	1	1
1	1	1	1	0	0	0
2	0		1	1	1	
3					0	
4					0	
5					1	
6						

In general, the steps are

1. Compute FB from the current message data bit and the state of the last shift register stage.
2. Use the new value of FB to compute the input to all stages that are fed by XOR gates.
3. The input to all other stages is simply the state of the preceding stage or, in the case of S1, the input is FB.
4. Immediately after a clock pulse, all stages will go to the state indicated by the input values computed in steps 2 and 3.

In this manner we can fill in the table until we have the values of S1, S2 and S3 following the sixth clock pulse:

CLOCK NUMBER	S1	S2_IN	S2	S3	MSG_DATA	FB
Start	0	1	0	0	1	1
1	1	1	1	0	0	0
2	0	0	1	1	1	0
3	0	1	0	1	0	1
4	1	1	1	0	0	0
5	0	0	1	1	1	0
6	0		0	1		

S3, S2, S1 now contain exactly the remainder that we obtained by long division in Section 10·3. At this point BCC ENABLE is asserted (see Figure 10–3), and some additional gating circuits (not shown in Figure 10–6) force FB to the 0 state. With FB held at 0 the CRC generator becomes an ordinary shift register, and the next three clock pulses put the BCC on the TRANSMIT DATA circuit (most significant bit first). It is also interesting to note that the bit values on FB (starting at the top of the table) are exactly the quotient obtained by the long-division method.

At the receiver we have an almost identical situation (see Figure 10–7). RE-CEIVE DATA and RECEIVE CLOCK are applied to a CRC generator representing the same polynomial used by the transmitter. If there are no errors, after the

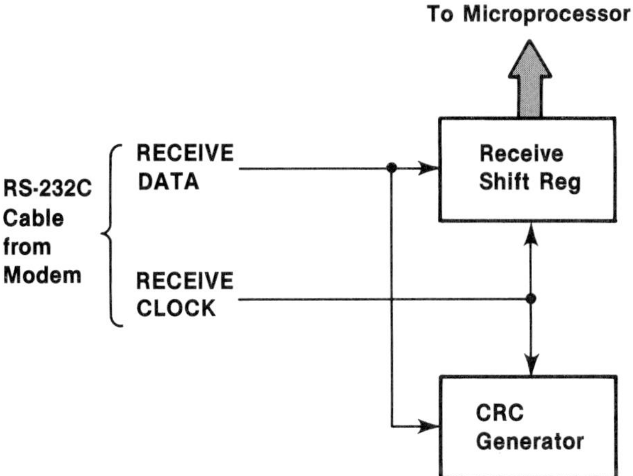

FIGURE 10–7   CRC Checking in a Receiver

last message bit is clocked in, the three stages of the receive CRC generator contain the exact bit pattern (100) produced by the transmitter. The next incoming bit on RECEIVE DATA (the first bit of the BCC) is equal to the value stored in S3, and receive's FB will therefore be 0 (see Figure 10–6). The same is true for the two remaining BCC bits. As a result, if there are no errors, after the BCC has been clocked in, all stages of the receive CRC generator contain 0. This is true regardless of the generator polynomial or the content of the data message. It makes error checking very simple from a hardware point of view.

Why bother with this procedure which, although it is relatively easy to implement in hardware, is conceptually so difficult? It has been demonstrated, both mathematically and in practice, that the CRC is far superior to the LRC, especially in the presence of burst errors. In actual communications systems, BCC_SIZE is typically 16 bits, occasionally more. Because of the feedback to intermediate stages, a value change in a single message bit will have a very complex effect on the contents of the shift register stages. Also, because of the length of the register, the effect of a single bit lingers for many clock pulses. This makes it unlikely that later errors will undo the effects of earlier ones as they did in the simple character parity case. The combination of these facts makes the probability of an undetected error extremely low when 16-bit CRC procedures are used. The CRC is so effective that protocols employing it often do not bother with parity bits on the individual characters. The two most commonly used generator polynomials are

$$CRC\_16 = X^{16} + X^{15} + X^2 + 1$$

and

$$CRC\_CCITT = X^{16} + X^{12} + X^5 + 1$$

10·5   ANOTHER VIEW OF PARITY

Let us take another look at the truth table for the XOR function:

A	B	A XOR B
0	0	0
0	1	1
1	0	1
1	1	0

If we look across the rows of the table, it is obvious that A XOR B can be interpreted as an even parity bit for the two input bits A and B. This is represented schematically in Figure 10–8A.

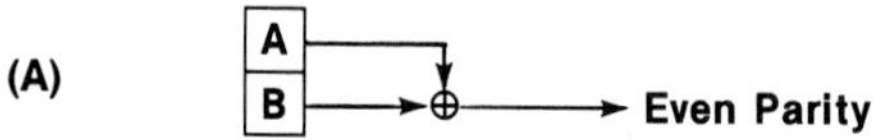

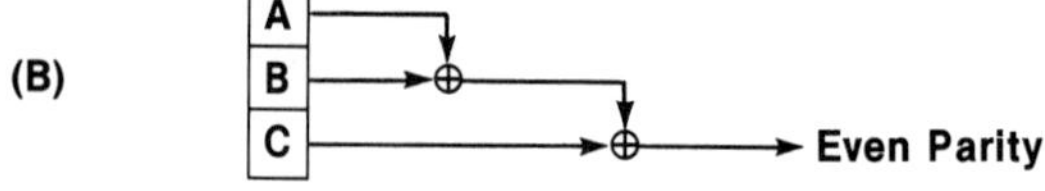

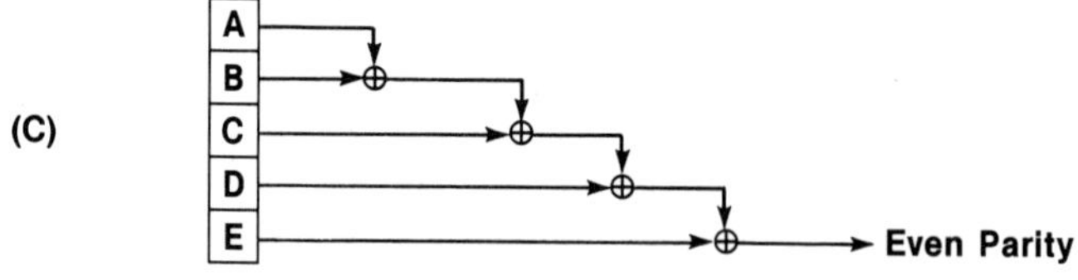

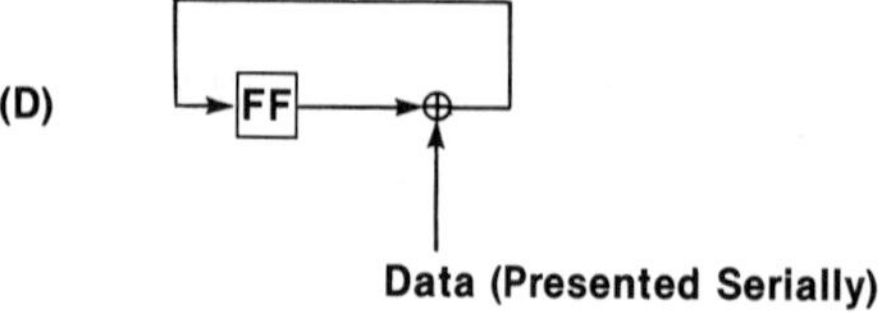

FIGURE 10–8   Parity Generators

Suppose we add one more bit (C); take the XOR of A and B, and use the result as the input to an XOR with C. This is shown in Figure 10-8B and the truth table is

A	B	C	(A XOR B) XOR C
0	0	0	0
0	0	1	1
0	1	0	1
0	1	1	0
1	0	0	1
1	0	1	0
1	1	0	0
1	1	1	1

Once again we can see from the table that the cascaded XOR is an even parity bit. We can extend this idea to any number of bits, using each additional bit as the input to an XOR with the result from all previous bits. A 5-bit case is illustrated in Figure 10–8C.

As the number of bits increases we find that we are using quite a few XOR gates. Perhaps we can save some hardware by the following scheme (see Figure 10–8D):

- Store the first data bit in a flip-flop.
- XOR the output of the flip-flip with the next data bit and store the result back into the same flip-flop. (This requires a clock which, once again, is not shown.)
- Repeat the previous step for all remaining data bits.

Since the flip-flop will always hold the cumulative XOR of all preceding data bits, we need only one flip-flop and one XOR gate to compute even parity. To properly store the first bit, the flip-flip must be initialized to 0 before the first bit is clocked in. After all bits have been clocked in, the flip-flip will contain even parity for the data.

Does this look familiar? What we have is a very small CRC generator for the polynomial $X + 1$. Parity on a serial bit stream is therefore a CRC procedure with BCC_SIZE = 1.

REVIEW QUESTIONS

1. In the following examples, the left column is odd character parity and the bottom row is an odd LRC. Are there any errors in the following messages? If so, show which control bits indicate their presence.

a.
```
0 0 1 0 1
0 1 1 0 1
1 0 0 1 1
1 1 1 0 0
‾‾‾‾‾‾‾‾‾
1 1 0 0 1
```

c.
```
1 0 0 1 1
0 0 1 0 0
1 1 1 0 0
0 1 0 1 1
‾‾‾‾‾‾‾‾‾
1 1 0 0 1
```

b.
```
1 0 0 1 1
0 0 1 0 0
1 1 1 0 0
0 1 1 0 1
‾‾‾‾‾‾‾‾‾
1 1 0 0 1
```

d.
```
1 0 0 1 1
0 1 1 0 0
1 0 1 0 0
0 1 1 0 1
‾‾‾‾‾‾‾‾‾
1 1 0 0 1
```

2. In a message containing 4 characters with 4 data bits in each character, find some error patterns that will not be detected by an LRC/character parity combination. Can you find any that involve less than 4 bits?

3. Use binary long division to compute the quotient and BCC for the following messages and divisors:

Divisor	Message (first bit is at left)
a. 1 1 0 1	1 1 0 1 0 1
b. 1 1 0 1	0 1 1 1 0 1
c. 1 0 1 0 1	1 0 0 0 0 1 0

4. Write the generator polynomial for each example in Review Question 3. Draw the CRC generator for each case and use the method of Section 10·5 to compute the BCC by applying the message and shift clocks to the generator circuit. Verify that the results match those obtained by long division in Review Question 3.

5. Draw the CRC generator/checker circuit for
   a. $X^8 + X^5 + X^2 + 1$
   b. $X^{12} + X^6 + 1$
   c. $X^8 + X^7 + X + 1$

chapter eleven

PROTOCOLS

The subject of protocols has been mentioned briefly at several places in the preceding chapters. A *protocol* is a set of rules designed to force the devices sharing a channel to follow orderly communications procedures. We now have enough background and vocabulary to examine this subject more thoroughly. The protocols we describe all involve blocked synchronous data formats. These protocols may be used for data transfers between two host computers or for transfers between a host computer and terminals. When terminals are involved, they are block mode devices, and there are usually several terminals attached to a multipoint line. It is this hardware complexity that creates the need for sophisticated protocols. While some aspects of protocols (flow control, for example) appear in conjunction with asynchronous devices, the typical asynchronous situation (a character mode terminal on a point-to-point link to a host computer) does not require complex control.

11·1 MAJOR FUNCTIONS OF PROTOCOLS

In this chapter we discuss the two most frequently encountered protocols for blocked synchronous data (BISYNC and HDLC). In terms of their detailed operation, they are very different, but they share a set of common goals. These are as follows:

FRAMING The *framing* function deals with the separation of message blocks from each other and the separation of individual blocks into *text* and *control* sections. Most protocols define a *maximum block size,* and a long *logical message* will have to be broken into blocks that do not exceed this limit. In addition to text, the block will usually contain control information such as the BCC or an *address field* (to identify the intended recipient on a multipoint circuit).

ERROR CONTROL In the previous chapter we discussed error detection. Both of the protocols we describe in this chapter use 16-bit CRC procedures. They also prescribe the course of action for recovery when an error is detected. In both cases retransmission requests are generated. The details of how these requests are handled are very different and have a strong impact on circuit throughput in the presence of errors.

LINE CONTROL *Line control* is a set of rules that specify which device on a channel has permission to transmit at any given time. On point-to-point full-duplex links this is obviously unnecessary, but it is critical on all multipoint circuits and on half-duplex point-to-point circuits.

FLOW CONTROL We have described this concept in earlier chapters and presented two implementations (XOFF/XON and secondary channel carrier). Blocked synchronous protocols use somewhat different procedures to accomplish the same function.

SEQUENCE CONTROL *Sequence control* is like error control, but it takes place at a higher level. In large data networks a message may pass through numerous telephone links and computer systems before it reaches its final destination. Complex hardware and software provide many opportunities for things to go wrong. Sequence control rules attempt to guarantee that message blocks will not be lost or duplicated and will arrive at their final destination in the order in which they were transmitted.

We now go on to describe the BISYNC and HDLC protocols. Aside from the many differences in operating detail, the primary distinction is that BISYNC is a half-duplex procedure and HDLC is a full-duplex procedure. These protocols are complex, and the examples we present are only typical and somewhat simplified. They are intended to give the reader the flavor of the protocols, not an in-depth description of their features.

11·2   HALF-DUPLEX PROTOCOLS (BISYNC)

The Binary Synchronous Communications (BISYNC) protocol was introduced by IBM in 1968. Its most significant feature is the half-duplex procedures it employs, even if full-duplex facilities are available. After every transmission (text or control) the sender must wait for a response before proceeding. A full-duplex channel can improve performance by eliminating turnaround delays, but BISYNC is still slow and inefficient when compared to newer protocols. Nevertheless, lots of BISYNC hardware and software was installed before the better protocols arrived, and BISYNC will remain part of the communications environment for years to come.

BISYNC data exchanges occur in two phases:

LINE CONTROL   No data is exchanged during this phase. Devices on the channel send various control messages and the result is a determination of which device will have control of the line during the next data transmission phase. As you might expect, line control procedures are different for point-to-point and multipoint circuits.

DATA TRANSMISSION   During this phase, the device that has gained "control" of the channel can send data messages to some other device on the circuit. The receiving device accepts the data and must acknowledge. No device other than the one momentarily in control can send data messages until after another line control phase takes place. In general, this cannot happen until the controlling device voluntarily gives up control. The reliable operation of protocols strongly requires that all devices on the channel play by the rules.

We will describe the data transmission phase first and then see how these procedures fit into both point-to-point and multipoint line control schemes.

11·3   BISYNC DATA TRANSMISSION (BLOCK FORMATS)

The basic format of a BISYNC data block is shown in Figure 11–1. The code may be ASCII or EBCDIC, but most BISYNC applications run on IBM hardware and EBCDIC is more typical. The components of the block are as follows:

Synchronization

As we stated in Chapter 6, bit synchronization for receiving synchronous data is provided by the RS-232C RECEIVE CLOCK from the modem. This still leaves us with the problem of character synchronization. All BISYNC transmissions are therefore preceded by at least two SYN (synchronization) control characters. When they are first turned on, BISYNC receiver circuits (the synchronous equivalent of the UART) are in what is called the *SYNC-hunt* mode. In this mode, after each RECEIVE CLOCK pulse, the receive circuits compare the contents of the Receive Shift Register to the binary pattern of the SYN character. Until a match is found, no data is delivered to the receiving software.

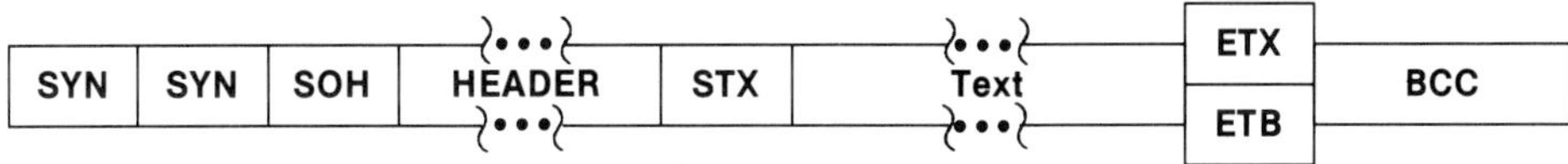

FIGURE 11–1   Format of a BISYNC Data Block

When a match is detected, we have achieved character synchronization. The receiver circuits exit SYNC-hunt mode and, in the new mode, deliver a character to the receive software after every eighth RECEIVE CLOCK pulse. (Blocked synchronous transmission virtually always uses an 8-bit character format.) At the end of the incoming block, the receiver circuits return to the SYNC-hunt mode in expectation of the start of the next block. On half-duplex channels, the carrier will be turned around between blocks. On full-duplex channels the interblock time will be filled with *pad* characters (usually steady MARK).

At least two SYN characters appear in front of all BISYNC transmissions (data or control). For the sake of brevity and clarity, we will not always show these in the following discussions of BISYNC.

Header

SOH (start of header) is a framing control character that introduces the variable length header field. This field is optional and may be absent. In situations where it is present, the header may contain any data the equipment designer wishes to include. It is usually some control information that does not logically fit into the text field, or some data that qualifies the meaning of the text field.

Text

This is another variable length field introduced by the framing control character STX (start of text). This field is compulsory in all data blocks. It carries the information content of the message. The text itself is usually a string of ASCII or EBCDIC characters.

The end of the text field is indicated by either ETB (end of text block) or ETX (end of text) framing characters. A logical message (for example, the full contents of a terminal screen) may be several times larger than the maximum block size. In this case, the message will be transmitted as a number of data blocks. The text field of all blocks except the last will be terminated by ETB. The last will be terminated by ETX. The BCC follows ETB or ETX.

While it is shifting in the text field, the receiver must scan the data for ETB or ETX to determine where the block ends. In most cases, the text field consists of printable characters and format effectors. None of these will ever match ETB or ETX, and the receiver will have no trouble finding the end of the block. Suppose, however, that we are transmitting the machine language version of a program or some other data that was composed of an unrestricted set of binary numbers. A byte in the middle of the text field may have the same binary value as ETB or ETX, causing the receiver to detect a false termination of the text field. Since it is highly unlikely that the 16 bits following the false ETB will be a correct BCC, we can never get this message through.

FIGURE 11–2   Block Format for Transparent Transmission

The solution is provided by a technique called *transparent text transmission*. In this case we use the slightly modified block format shown in Figure 11–2 (without the optional header field). Instead of STX we have the two character sequence DLE STX at the beginning of the text field. (DLE is "data link escape.") This tells the receiver to obey the following rules for transparent transmission:

- If the binary value of a character in the text field does not match DLE, it is not considered as a possible control character. These characters are delivered to the software as text characters regardless of their binary value.
- If a received character does match DLE, discard it and check the following character to see if it is ETB or ETX. If this is true, we have found the end of the block, and the BCC is checked. If the following character is not ETB or ETX, deliver it to the software as an ordinary data character.

Note that if the transmitter wants to send DLE as part of a transparent message, it must actually send DLE DLE since the second rule tells the receiver to discard the first DLE it finds.

11·4   BISYNC DATA TRANSMISSION (RESPONSE MESSAGES)

The BISYNC data blocks described earlier can only be transmitted by the device that has gained control of the channel. The transmissions are always sent to a particular destination device. (Exactly how we determine which device is the destination will be described later in the chapter.) The destination device is obliged to respond with one of the three following messages:

1. NAK (NEGATIVE ACKNOWLEDGE). NAK is sent in response to a message received with a BCC error. The last data block will be retransmitted.
2. ACK 0 (ACKNOWLEDGE 0). This is actually a 2-character sequence (DLE 0 in ASCII) transmitted in response to an even numbered data block (second, fourth, . . .) that was received without errors.
3. ACK 1 (ACKNOWLEDGE 1). 2-character sequence (DLE 1 in ASCII) transmitted in response to an odd numbered data block that was received without errors.

These response messages are always preceded by a few SYN characters, but they do not carry a BCC. A typical transaction is illustrated in Figure 11–3. West has obtained control of the line and wants to transmit a 3-block message. (We have not bothered to show SYN, Header, or BCC.) The first block is sent with ETB as a terminator and received correctly. The response is therefore ACK 1. Block 2 is hit by an error, and the response is NAK which causes a retransmission. Block 2 arrives intact on the second try, and the response is ACK 0. Block 3 is the last and is therefore terminated by ETX. It also arrives without errors, and the response goes back as ACK 1.

The alternation of ACK responses is a form of sequence control. If two consecutive ACK 0 (or ACK 1) responses were received, it would be an indication that a block had been lost.

What factors affect our choice of block size? Any error, even a single bit, requires that an entire block be retransmitted. On this basis we tend to keep the blocks small, because larger blocks have a higher probability of being hit by an error and take longer to retransmit. On the other hand, if blocks are extremely small, the non-information-carrying "overhead" (SYNs, framing char-

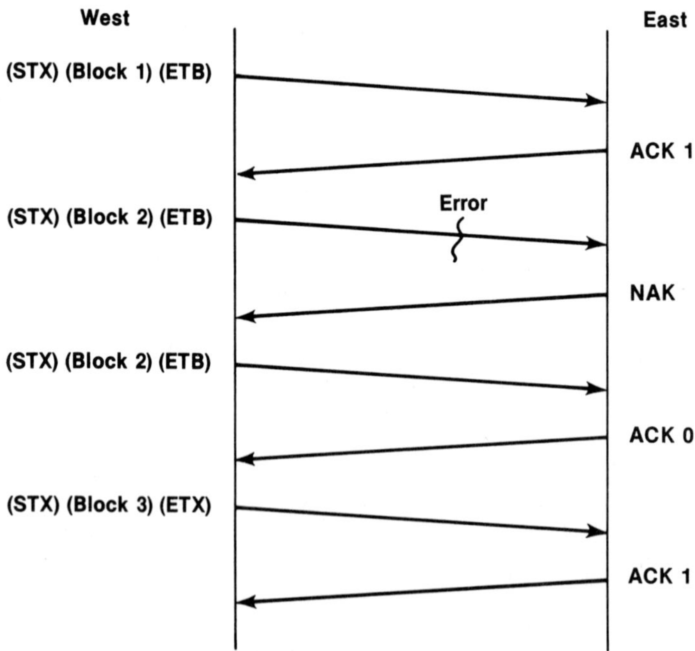

FIGURE 11–3 BISYNC Data Exchange

FIGURE 11–4   Point-to-Point BISYNC Connection

acters, BCC) becomes a significant fraction of total line time. Between these two extremes we can calculate an optimum block size. For most protocols and circuit conditions this optimum turns out to be fairly close to 256 bytes.

11·5   BISYNC LINE CONTROL (POINT-TO-POINT)

Point-to-point BISYNC is typically used to connect two host computers. Traffic on these connections usually consists of large data files being transferred in one direction or the other. This is very different than terminal/host-computer traffic, which consists of many short messages with rapid changes of direction.

A point-to-point configuration is illustrated in Figure 11–4. Since both ends of the line generate roughly equal traffic volumes, the host computers are considered equal contenders for control of the line. In fact, the line control procedure used on these links is frequently called *contention*. It operates as follows:

- When the line is free, either end can *bid* for control.
- If the other end has nothing to send, it accepts the bid, and the bidder is given control of the line.
- Once a device has gained control it enters the data transmission phase described in the previous section.
- When the controlling device has completed its transmission, it releases the line. The line is once again open for bids from either side.
- If there is a *bid collision,* one side backs down and allows the other side to take control. The side that backs down is selected by prearranged agreement between the people operating the host computers.

A typical BISYNC point-to-point exchange is illustrated in Figure 11–5. Starting from the free line state, West bids for the line with a single character ENQ (enquire) message. East accepts by sending ACK 0, and we enter the data transmission phase with West in control. Notice the ACK alternation by East (ACK 0 to accept the bid; ACK 1 for the first message; etc.) Data transfer continues until West releases the line with the single character EOT (end of transmission) message. The line is free once again. As usual, ENQ and EOT are preceded by a few SYNs. Neither one carries a BCC.

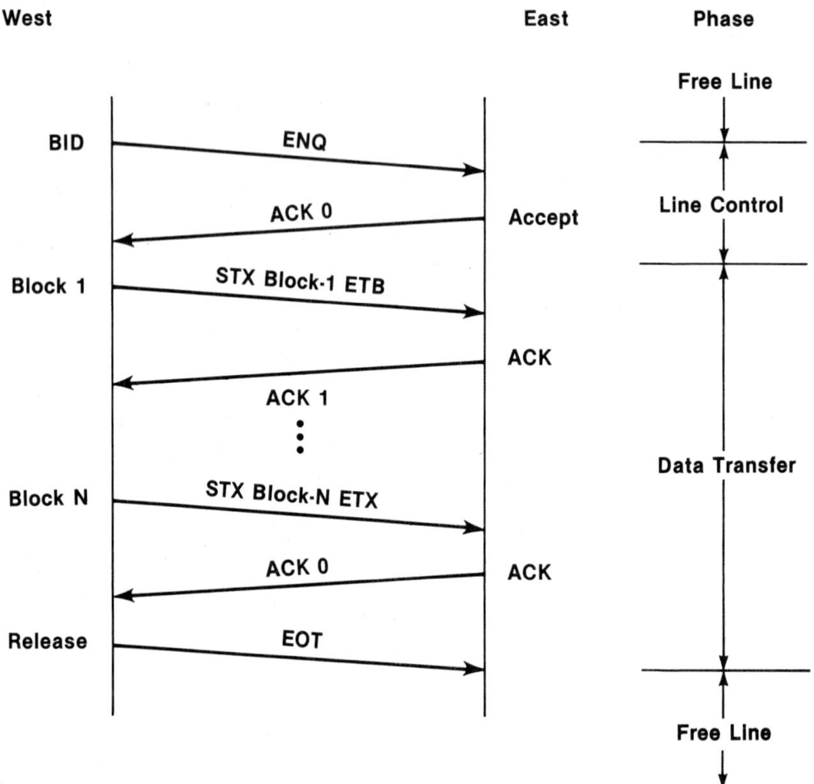

FIGURE 11–5   Point-to-Point BISYNC Traffic

11·6   BISYNC LINE CONTROL (MULTIPOINT)

On multipoint BISYNC circuits (see Figure 11–6) we have a number of block mode terminals (called *tributaries)* and the host computer (called the *supervisor*). In this situation the traffic is definitely unbalanced. Each terminal generates an occasional message. The host computer must respond to all terminals. The host computer (supervisor) therefore has complete command of the line. Tributaries can control a data transmission phase only when given permission by the host computer.

A block mode terminal will have a message to send whenever its operator hits the ENTER key. On a BISYNC multipoint circuit, the terminal cannot send the message immediately. It must wait until it is given permission to transmit by the supervisor. Permission is granted through a process called *polling,* which operates as follows:

• The supervisor periodically sends an interrogation message (called a *specific poll*) to each terminal. Each terminal has a unique *poll address* which enables it to recognize its own poll message.

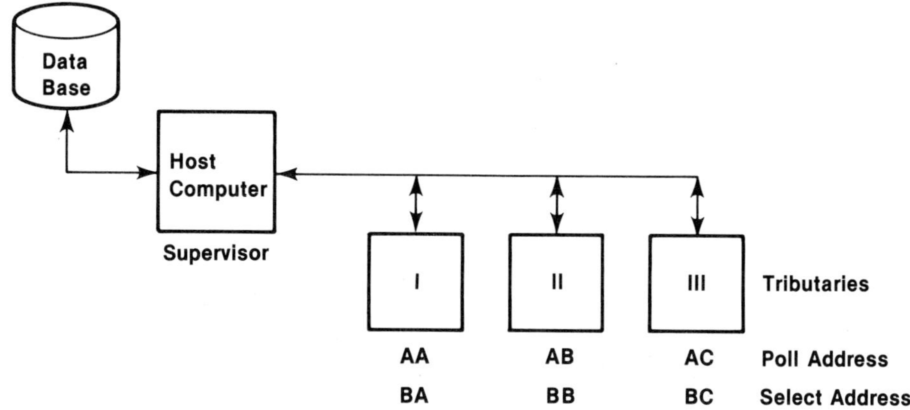

FIGURE 11–6   Multipoint BISYNC Connection

- If the terminal has no data to send it responds to the poll with EOT.
- If the terminal has a message pending it enters a data tansmission phase, as described in Section 11·4, and controls the line until it has emptied its message buffer. The host computer responds with ACK 0/1 or NAK while the terminal is transmitting its message blocks.
- The host computer then polls the next terminal on its list. When it reaches the end of the list it returns to the top. This process is sometimes called *roll call polling*.

Polling takes care of data transmissions from the terminals to the host computer. Data flow in the other direction is handled by a process called *selection*. When the host computer has a message for a terminal it inserts a selection sequence between polls. Selection operates as follows:

- The host computer sends a selection command containing a specific terminal's *select address*.
- If the addressed terminal is ready to accept the data from the host computer, it responds with ACK 0, and we enter a data transmission phase with the host computer controlling the line. The terminal responds with ACK 0/1 or NAK until the message has been completely transmitted.
- If the terminal is busy it can flow control the host computer by responding with a WACK (2-character sequence meaning "wait before transmit". The host computer responds to WACK with EOT and attempts the selection again at a later time. Typical BISYNC terminals produce a WACK response only if they are making use of an auxiliary printer that cannot accept data at the line rate.

In Figure 11–6 we have labeled the terminals I, II, and III. The poll and select addresses (which are typically 2-character sequences) are shown below each terminal. A data base application is a typical use of such a configuration. Terminals generate information requests, and the data base program in the host computer formulates answers. Since this involves disk access, there is a delay before the answer is ready. During this time the host computer continues polling. When the answer is ready it is returned to the terminal by a selection process. Figures 11–7 illustrates such an exchange on a multipoint BISYNC line.

The host computer polls Terminal I by sending its poll address (AA) followed by ENQ. Terminal I responds with a 1-block message requesting information from the data base. While the data base program is formulating the answer, the host computer polls Terminal II (poll address AB). Terminal II has no traffic and responds with EOT.

A 2-block response is now ready for Terminal I. The host computer therefore selects Terminal I by sending its select address (BA) followed by ENQ. Terminal I is not busy and responds with ACK 0. The host computer then sends the 2-block message, which is received without errors. Transmission to Terminal I is now complete, and the host computer returns to polling. After it completes its list by polling Terminal III, it loops back to the top of the list again.

To get a feel for the time involved in BISYNC data transmission sequences, let us compute the duration of the first exchange in Figure 11–7, subject to the following assumptions:

- characters are 8 bits long;
- each transmission is preceded by two SYN characters;
- the data field in the second block consists of 10 characters;
- the line is 2400 bits/s half-duplex;
- the turnaround time is 150 ms.

We must compute the sum of the following times:

A. Poll I
B. line turnaround
C. Information request message from Terminal I
D. line turnaround
E. ACK from the host computer

The character time is

$$\frac{\text{BITS\_PER\_CHAR}}{\text{B\_P\_S}} = \frac{8}{2400} = 3.3 \text{ ms.}$$

We therefore have

A: SYN SYN A A ENQ $= 5$ chars
$$5 * 3.3 = 16.5 \text{ ms}$$
B: turnaround $= 150$ ms

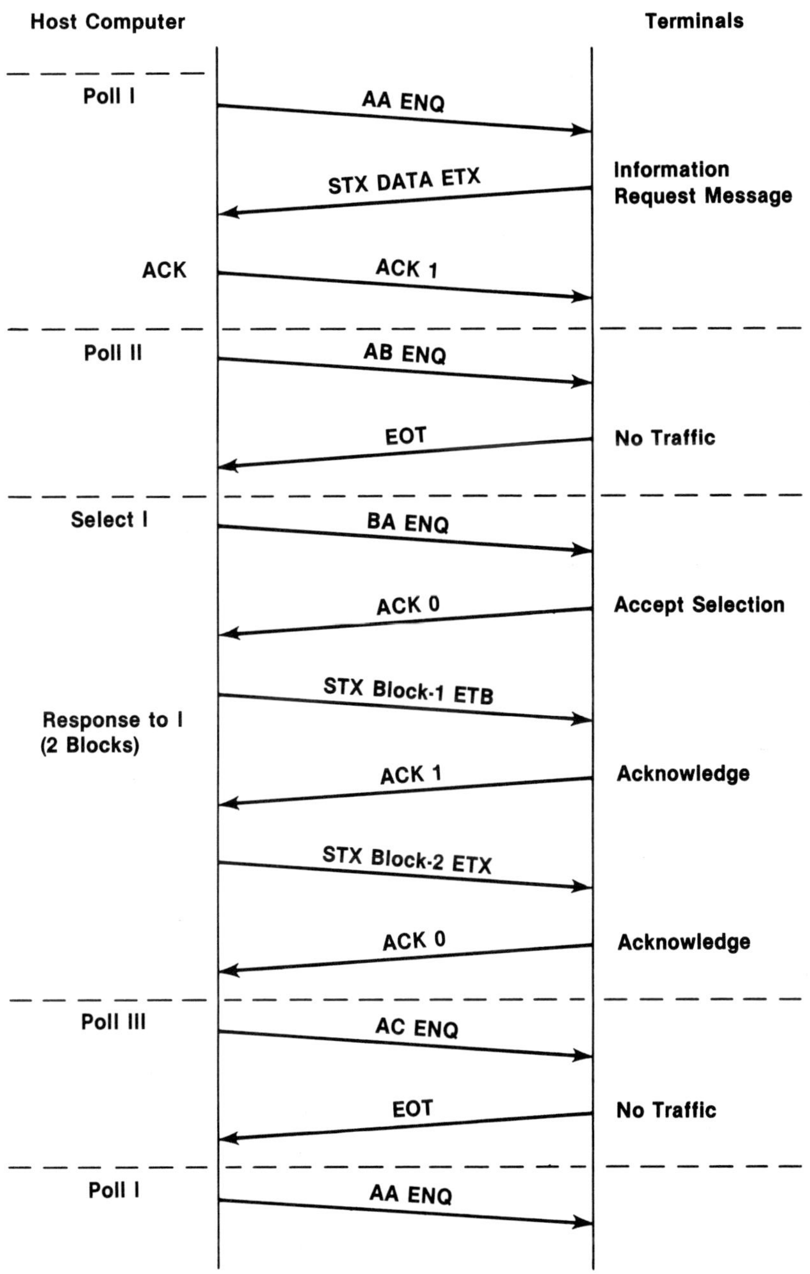

FIGURE 11–7　Multipoint BISYNC Traffic

C: SYN SYN STX (10 data chars) ETX (2-char BCC) = 16 chars
$$16 * 3.3 = 52.8 \text{ ms}$$
D: turnaround = 150 ms
E: SYN SYN (2-char ACK 1) = 4 chars
$$4 * 3.3 = 13.2 \text{ ms}$$

Therefore,

$$A + B + C + D + E = 16.5 + 150 + 52.8 + 150 + 13.2$$
$$= 383 \text{ ms}$$

The most significant feature of this process is its half-duplex nature. Notice that all polling (and therefore all data transmission from any terminal) had to stop and wait until the selection of Terminal I was complete. This will happen in BISYNC even if the modems and the telephone lines are full-duplex. In the next few sections we will see how the HDLC protocol can take advantage of full-duplex facilities by allowing the host computer to simultaneously transmit to one terminal while accepting data from another.

11·7   FULL DUPLEX PROTOCOLS (HDLC)

HDLC (High-level Data Link Control) is a full-duplex protocol whose definition was published by the International Standards Organization (ISO) in 1977. It very closely resembles a protocol, called SDLC (Synchronous Data Link Control), introduced by IBM in 1974. At the level of detail appropriate for this book, we may consider HDLC and SDLC to be two different names for the same protocol. Since HDLC bears the stamp of international authority, we will use that name. While HDLC can operate on a half-duplex link, its real powers are apparent when a full-duplex facility is available. Our examples illustrate both of these operating modes.

HDLC has a number of different operating modes (known as *classes of procedure*). We will primarily examine the multipoint terminal/host-computer configuration which gives us a basis for direct comparison with the BISYNC procedures described in the previous section. The formal name for this class of procedure is *normal response mode*, and the block diagram in Figure 11–6 applies. In HDLC, the host computer is called the *primary station;* the terminals, *secondary stations*. The distinguishing feature of normal response mode is the fact that a secondary station cannot initiate a transmission unless given permission to do so by the primary. The exact procedure involved will be explained shortly.

As was the case with BISYNC, the intent of this chapter is to convey an understanding of the nature of the protocol, not to present a design handbook. Hence, the examples are typical and slightly simplified.

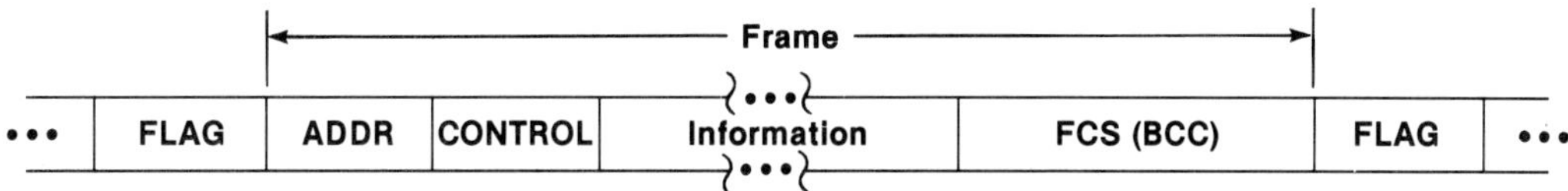

FIGURE 11–8   HDLC Frame Structure

11·8   HDLC FRAME STRUCTURE

The primary transmission unit of HDLC is the frame illustrated in Figure 11–8. An HDLC frame is functionally equivalent to a block in BISYNC. The components of the frame are described as follows.

Flag

The 8-bit *flag* pattern (binary 0111 1110) indicates the beginning and end of frames and fills all interframe line time. It is the only framing character in HDLC. As was the case in BISYNC, we have a transparency problem. An HDLC transmitter must guarantee that a flag pattern never occurs within the limits of a frame. Even the case where the flag pattern spans the boundary between two characters is not allowed. The technique used to achieve transparent transmission in HDLC is called *zero stuffing*. It operates as follows (see Figure 11–9 for an illustration):

• HDLC flags are usually generated by an LSI transmitter circuit (roughly equivalent to a UART) which sends flags to the modem whenever it is not

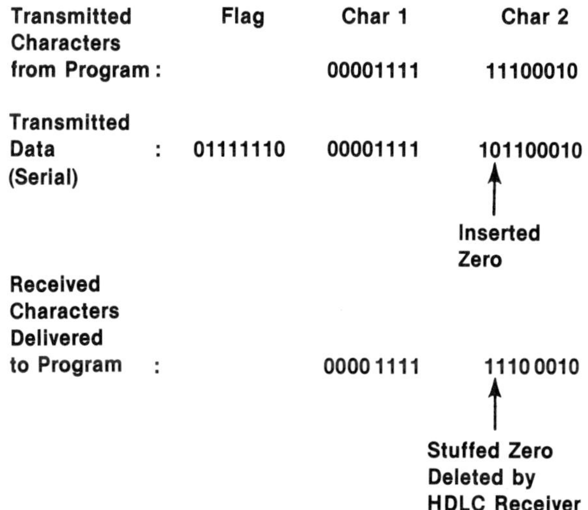

FIGURE 11–9   HDLC Zero Stuffing

receiving parallel character transfers from the program. The primary identifying characteristic of the flag is six consecutive 1s.

- If the transmitter circuit detects five consecutive 1s in frame data it is getting from the program, it automatically inserts an extra 0 into the serial bit
stream. In Figure 11–9 this occurs just after the first bit of char 2 because of
the four 1s at the end of char 1.
- At the other end of the link, an LSI receiver circuit is also monitoring the
serial bit stream for five consecutive 1s. When five 1s are detected the receiver checks the state of the next bit. If it is a 0, the receiver assumes that
this bit was "stuffed." The stuffed zero is discarded, and the receiver continues assembling parallel characters for delivery to the program.
- When the receiver detects six consecutive incoming 1s, it assumes that the
end of the frame is at hand. Zero stuffing by the transmitter would make the
occurrence of this pattern impossible anywhere within the frame.

Following the last interframe flag, are two fields, called *address* and *control*. These are always exactly 8 bits long.

Address Byte

As it did in BISYNC, the address byte is used to uniquely identify a secondary station on the multipoint circuit. HDLC does not make a distinction between poll and select addresses. The 8-bit size of the HDLC address allows up
to 256 secondary stations to be connected to the multipoint circuit.

For use in later examples, we define new HDLC addresses for the terminals
in Figure 11–6:

Secondary Station	HDLC Address
Terminal I	1
Terminal II	2
Terminal III	3

The primary station does not have an address:

- Frames transmitted by the primary station contain the address of the secondary that is the intended destination of the frame.
- Frames transmitted by a secondary station contain the transmitting station's
own address.

Control Byte

The *control* byte indicates the type of frame being transmitted and is the
key to understanding the HDLC protocol. It is broken down into several sub-
fields, as indicated in Figure 11–10. A frame can belong to one of three major
categories, as indicated by bits 0 and 1 of the control byte:

- *Information transfer (I) frames* (Figure 11–10A): These are the basic mechanism for moving data in HDLC.
- *Supervisory frames* (Figure 11–10B): These provide the mechanism that controls data movement in HDLC. The control byte of supervisory frames carries a 2-bit *type* field that further classifies these frames into
    - •• RECEIVE READY (RR): handy little frames that are used for polling, acknowledgment, and error control;
    - •• RECEIVE NOT READY (RNR): used for flow control; RNRs acknowledge received frames, but notify the sender that the receiver has become busy;
    - •• REJECT (REJ): used for sequence and error control.
- *Unnumbered frames* (Figure 11–10C): Frames used for initialization, setting modes, and placing stations off-line or on-line. We will not discuss these in detail.

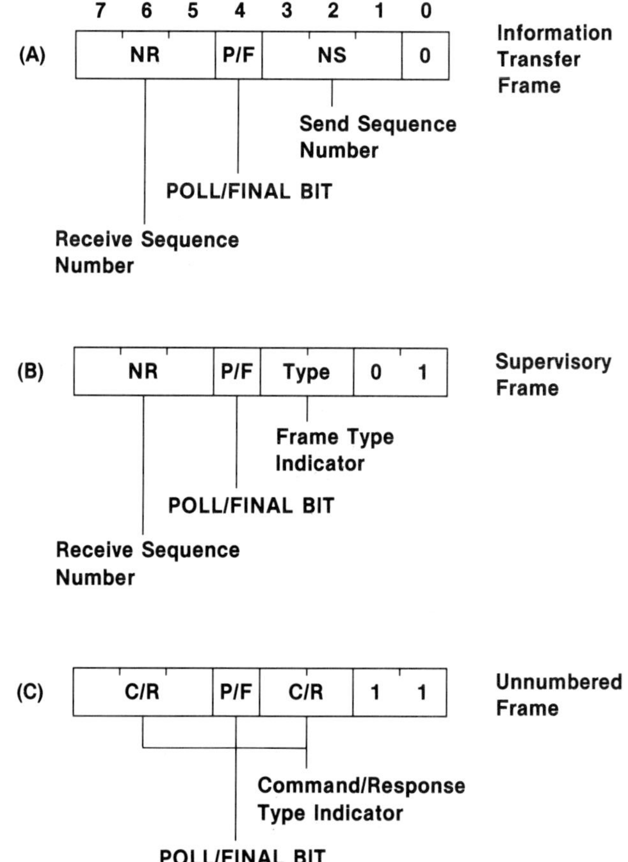

FIGURE 11–10   HDLC Control Field Structure

If this seems a little confusing, we will soon see some examples that will clarify the situation. First we must acquire a little more HDLC vocabulary.

For all frame types, the control byte carries a *poll/final* (P/F) bit. If this bit is set, its meaning depends on the source of the frame:

Poll
In a frame transmitted by a primary station, P/F = 1 means that this frame is a poll. The secondary station addressed by this frame must respond to the primary.

Final
In HDLC, a station may send multiple frames without waiting for an immediate acknowledge. In a frame transmitted by a secondary station, P/F = 1 means that this is the last frame of a transmission sequence.

If P/F = 0, the frame has neither the poll or the final meaning.

In information transfer frames, the control byte contains a 3-bit *send sequence counter* (NS). NS is initialized to 0. Each time a station transmits an information transfer frame, the value of NS is incremented by 1 from the value that the station used for the previous information transfer frame. After NS reaches 7 (the maximum value a 3-bit field can hold), it is reset to 0 on the next information transfer frame. The receiver uses the number for sequence control.

In information transfer and supervisory frames, the control byte carries a 3-bit *receive sequence number* (NR). NR indicates the value that a station expects to see in the NS field of the next information transfer frame it receives. It is used as an acknowledgment. To restate the definition:

If a station sends a frame whose control byte contains NR = 5, this means that it has received, without errors, all information transfer frames up to and including the one that carried NS = 4. This station now expects the next incoming information transfer frame to carry NS = 5.

Returning to Figure 11–8, there are two parts of the frame we have not yet discussed.

Information Field

This field contains the actual data carried by the HDLC frame. The field is variable length and only appears in information transfer frames. The HDLC protocol does not specify a maximum size for this field, but most implementations place an upper limit on total frame size. Block size considerations similar to those described for BISYNC apply here too.

Frame Check Sequence (FCS)

This is a 16-bit BCC computed from the CRC_CCITT polynomial. Notice that there is no framing character separating the BCC from the end of the *information field*. In the receive circuit, the input to the CRC *checker* must be delayed 8 bits from the input to the *flag detector*. When the terminating flag is detected, the CRC checker may then be immediately frozen at the exact end of the BCC.

11·9 TYPICAL HDLC FRAME SEQUENCES

Figures 11–11, 11–12, and 11–13 show some typical data exchanges on a multipoint HDLC circuit. The notation we use to describe the frame is

FRAME_TYPE   ADDRESS   (P/F)   NR_VALUE   NS_VALUE

Step	Primary	Secondaries	Comment
1)	RR A=1 (P) NR=0	→	Poll, PRI → Term I
2)		← I A=1 NR=0 NS=0	Info, Term I → PRI
3)		← I A=1 NR=0 NS=1	
4)		← I A=1 NR=0 NS=2	
5)		← I A=1 (F) NR=0 NS=3	
6)	RR A=1 NR=4	→	ACK, PRI → Term I
7)	RR A=2 (P) NR=0	→	Poll, PRI → Term II
8)		← RR A=2 (F) NR=0	No Traffic, Term II → PRI
9)	RR A=3 (P) NR=0	→	Poll, PRI → Term III
10)		← I A=3 NR=0 NS=0	Info, Term III → PRI
11)		← I A=3 (F) NR=0 NS=1	
12)	I A=3 NR=2 NS=0	→	Info & ACK, PRI → Term III
13)	I A=3 NR=2 NS=1	→	Info, PRI → Term III
14)	I A=3 (P) NR=2 NS=2	→	Info & Poll, PRI → Term III
15)		← RR A=3 (F) NR=3	ACK, Term III → PRI
16)	RR A=1 (P) NR=4	→	Poll, PRI → Term I

FIGURE 11–11 Half-Duplex HDLC

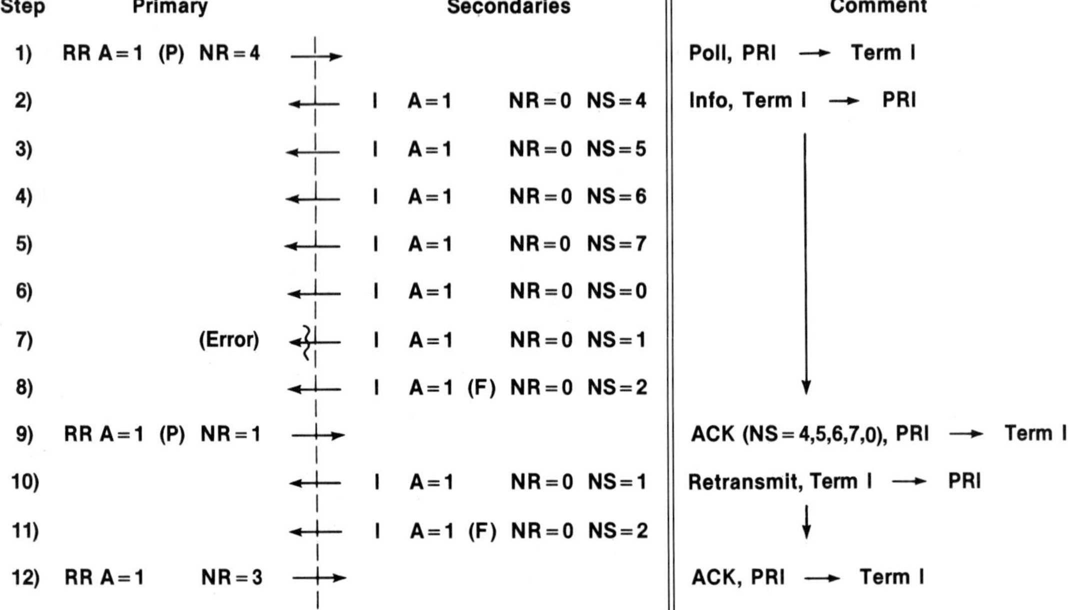

FIGURE 11–12   HDLC Error Control

Step	Primary		Secondaries	Comment
1)	RR A=2 (P) NR=0	→		Poll, PRI → Term II
2)		←	I A=2 NR=0 NS=0	Info, Term II → PRI
3)		←	I A=2 NR=0 NS=1	Info, Term II → PRI
4)	I A=1 NR=3 NS=0	→		Info, PRI → Term I
5)		←	I A=2 NR=0 NS=2	Info, Term II → PRI
6)	I A=1 NR=3 NS=1	→		Info, PRI → Term I
7)		←	I A=2 (F) NR=0 NS=3	Info, Term II → PRI
8)	I A=1 NR=3 NS=2	→		Info, PRI → Term I
9)	RR A=2 NR=4	→		ACK, PRI → Term II
10)	I A=1 (P) NR=3 NS=3	→		Info & Poll, PRI → Term I
11)		←	RR A=1 (F) NR=4	ACK, Term I → PRI

FIGURE 11–13   Full-Duplex HDLC

If a frame type is not information transfer, NS_VALUE is not present. If P/F is not set, this field is not shown. If P/F is set, we indicate its source by (P) or (F). In our examples FRAME_TYPE can be

I = Information Transfer
RR = Receive Ready

The block diagram for all examples will be Figure 11–6. We assume that the first example begins in the initialized state. That is, the first information transfer frame sent by each device will have NS = 0. As we will see in the examples, the primary station maintains independent sequence counts with each secondary station.

Half-duplex HDLC

Figure 11–11 illustrates a half-duplex HDLC sequence. At Step 1, the primary polls Terminal I by specifying its address (A = 1) in an RR frame with P set. NR = 0 in this frame indicates that the primary is expecting the next information transfer frame from Terminal I to have NS = 0. Terminal I does, in fact, have a message which is transmitted to the primary in Step 2 through Step 5. Note the following features of these four information transfer frames:

- Terminal I transmits several consecutive frames without waiting for an acknowledgment from the primary for each one.
- In each frame, NS is incremented by 1 from its value in the previous frame.
- Since Terminal I has not yet received any information transfer frames from the primary, NR = 0, and it remains there for all four steps.
- In Step 5, Terminal I turns on the F bit to indicate that this is the end of the current sequence.

At Step 6, the primary acknowledges all these frames with a single RR with NR = 4. This states that the primary has received, without errors, all information transfer frames from Terminal I up to, and including, the one with NS = 3 (the last one). The next information transfer frame that the primary expects from Terminal I should carry NS = 4. However, since the RR in Step 6 did not have the P bit set, Terminal I cannot send any more information at this time.

At Step 7, the primary polls Terminal II. This is identical to Step 1 except for the address field. Terminal II has no pending information, and it sends the HDLC "no traffic" response (Step 8): an RR frame containing its own address (A = 2) with the F bit set.

In Step 9, the primary polls Terminal III. Terminal III has two frames of information which it sends in Steps 10 and 11. The primary now owes Terminal III an acknowledgment for frames with NS = 0 and NS = 1. This requires that the primary send Terminal III a frame with NR = 2. We had a similar situation in Step 6. There the acknowledgment was carried by an RR. At Step 12, however, the primary has some information to send to Terminal III as well as an acknowledgment. In HDLC, it is possible to combine these two functions because an information transfer frame carries the NR field. We can see this in Step 12, where the primary sends an information transfer frame (NS = 0) that also carries NR = 2. This is called a *piggybacked acknowledgment.*

The primary sends two more frames of information to Terminal III (Steps 13 and 14) and, in Step 14 also demands an acknowledgment by setting the P bit. In Step 15 Terminal III acknowledges frames 0, 1, and 2 and from the primary by sending RR with NR = 3.

Even for this half-duplex example, the advantages of HDLC over BISYNC are apparent:

- Since several frames can be transmitted without waiting for an acknowledgment for each one, HDLC eliminates many of the line turnaround delays and small control messages that would be necessary in BISYNC.
- We gain even more efficiency by the ability to combine an acknowledgment with an information frame.

HDLC Error Control

We have not yet seen how errors are handled in HDLC. In Figure 11–12, the primary polls Terminal I (Step 1), expecting a frame with NS =4. Terminal I sends seven consecutive information transfer frames (steps 2–8). Notice how NS increments up to 7 and then rolls back to 0.

Assume that there was a transmission error at Step 7 (NS = 1). The primary notes the sequence number (NS) of the frame it was expecting at the time the error occurred, but takes no other action at that time. If all seven frames from Terminal I had been received without errors, the appropriate response from the primary would be a frame with NR = 3 (acknowledging received frames with NS up to 2). At Step 9, however, the primary sends RR with NR = 1, indicating that the last valid frame received carried NS = 0. The RR in Step 9 also has P set, demanding a response from Terminal I. The appropriate response from Terminal I is the retransmission of the frame the primary claims it is expecting plus all subsequent frames. In Steps 10 and 11 Terminal I retransmits the frame that was hit by the error and the following one (NS = 1 and NS = 2). HDLC requires that no information frames are transmitted out of sequence. The frame with NS = 2 must therefore be retransmitted in spite of the fact that it was received correctly. If the error had occurred at Step 4 (NS = 6), the RR at Step 9 would have had NR = 6 and Terminal I would have retransmitted five frames.

In this, and the preceding example, we have illustrated cases where an HDLC station sends several consecutive information frames without waiting for an acknowledgment. These temporarily unacknowledged frames are called *outstanding frames.* Is there an upper limit to the number of outstanding frames? Yes, the limit is 7. We can deduce the reason from Figure 11–12. Suppose we had allowed Terminal I to send one more information transfer frame following Step 8. This frame would carry NS = 3, and, if everything had been received without error, the appropriate response from the primary would be a frame with NR = 4. We now have introduced a confusion factor. Does the NR = 4 from the primary mean

- an acknowledgment of eight frames received without error?
OR
- an error was detected at Step 2 (NS = 4) and all eight frames must be retransmitted?

To avoid this problem we must limit the maximum number of outstanding frames to seven.

Several other error recovery methods are available in HDLC:

- timeouts and retries of polls to nonresponding stations;
- use of the REJECT supervisory frame in response to sequence count errors in NS;
- use of some unnumbered frame types to reject transmissions that grossly violate the rules of HDLC.

We will not discuss these mechanisms in detail.

HDLC stations also implement flow control by means of the RNR (RECEIVE NOT READY) supervisory frame. RNR indicates that a station has become busy and cannot accept any more information transfer frames. Since RNR frames carry an NR field, they can acknowledge the last frame received before the busy condition occurred. When the station is no longer busy it responds with RR.

Full-duplex HDLC

In Figure 11–13 we see an HDLC example that starts with the primary station polling Terminal II. Terminal II responds with a sequence of information frames which, until we get to Step 3, closely resembles the earlier examples. Suppose that at some earlier time Terminal I had transmitted a data base request to the host computer. The data base program required some time to process this request, and a 4-frame response is now ready for Terminal I. In our multipoint BISYNC example (see Figure 11–7) a similar response was sent by means of a select sequence. The select, however, could not be initiated while a poll sequence was in progress. In full-duplex HDLC there is no need to wait.

The primary can begin transmission of the response to Terminal I (Step 4) the instant it becomes available. In Figure 11–13 the beginning of Step 4 occurs before the end of Step 3. There will be a similar overlap in time between Steps 5 and 6 and, once again, between Steps 7 and 8. Step 7 was the last input frame from Terminal II (notice the presence of the F bit), and the primary therefore owes Terminal II an acknowledgment. The primary's HDLC transmitter inserts this acknowledgment (Step 9) between the third and fourth information frames sent to Terminal I. This simultaneous communication between the host computer and two different terminals would be impossible under BISYNC or any other half-duplex protocol.

11·10   THE DESIGN OF LARGE COMMUNICATIONS CONTROLLERS

We have said very little about the way HDLC transmission controllers are actually constructed. In general, they are complex combinations of hardware and software, and a book could easily be devoted to this subject alone. Their functions are usually separated more or less as follows:

Hardware
- physical line interface (voltage conversion)
- flag generation
- zero stuffing
- BCC generation
- error detection
- parallel-to-serial conversion

Software
- error recovery
- polling and acknowledgment
- NS and NR counter management
- measuring traffic flow statistics
- providing operator interface functions (bringing stations on line, changing device addresses, reporting error conditions, etc.)
- breaking long logical messages into a number of smaller frames (and vice versa)

These logical messages are transferred between the HDLC software and other programs (such as the data base program) which are time-sharing the host computer with the HDLC software. When many terminals are present (in some installations there are hundreds), complex communications functions such a BISYNC or HDLC consume so much host-computer time that other programs begin to slow down. To alleviate this problem some manufacturers have isolated the communications functions and installed them in a separate device called a *communications front end* (CFE), which is illustrated in Figure 11–14. The CFE is itself a fairly large computer. Logical messages are transferred between the CFE and the host computer via a high-speed bus.

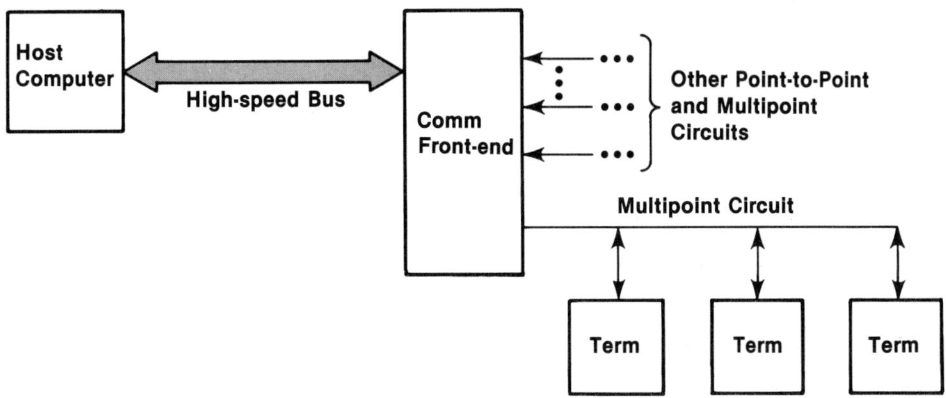

FIGURE 11–14   Communications Front End

REVIEW QUESTIONS

1. The following sequence of ASCII control characters must be transmitted as text on a ASCII BISYNC circuit in a single block.

   DC1   LF   ETX   ACK   DLE   SOH   NAK

   Draw the block, in transparent format, that will correctly carry this data. (Do not bother with the header field.)

2. Look at the first five BISYNC transmission blocks in Figure 11–7 and make the following assumptions:
   - characters are 8 bits long;
   - each transmission is preceded by two SYN characters;
   - the data field in the second transmission consists of 32 characters;
   - the line is 4800 bits/s half-duplex;
   - the turnaround time is 150 ms.
   a. What is the total number of characters transmitted?
   b. What percentage of these are data?
   c. How much time is consumed by this entire sequence?
   d. What percentage of the total time is represented by data?
   e. What percentage of the total time would data represent if a full-duplex circuit was available?

3. Using the same assumptions (half-duplex) given in Review Question 2, how long will it take the host computer to complete a polling cycle of all three terminals on the circuit if all respond with "no traffic"?

4. The following bit patterns appear in the middle of an HDLC frame (from the program's point of view):

Char 1	Char 2	Char 3
01011111	00100011	11111000

How will this data appear after it is placed on the line by an HDLC transmitter? Indicate the position of any stuffed zeros.

5. What is the binary pattern of the control field in the following HDLC frames? (For supervisory frames, the frame type indicator is encoded as follows: RR = 00; RNR = 10; REJ = 01)

a. RR   (P)   NR = 3
b. I   NR = 2   NS = 7
c. RNR   (F)   NR = 6
d. REJ   NR = 1

6. On a multipoint HDLC circuit, the primary sends the following frame to a secondary whose address is 4:

$$I \quad A = 4 \quad (P) \quad NR = 1 \quad NS = 4$$

The secondary has no further data to send and responds with RR. In the correct response:
a. What is the value in the address field?
b. What is the state of the P/F bit?
c. What is the value of NR?

7. The following is a sequence of frames from an HDLC primary:

$$I \qquad A = 2 \quad NR = 2 \quad NS = 3$$
$$I \qquad A = 2 \quad NR = ? \quad NS = ?$$
$$I \quad (P) \quad A = 2 \quad NR = 2 \quad NS = 5$$

a. What are the correct values of NR and NS in the second frame?
b. If the secondary responded to this sequence with

$$RR \quad (F) \quad A = 2 \quad NR = 4$$

what action should the primary take?
8. Redraw Step 1 through Step 6 of Figure 11–11 as a BISYNC exchange. Assuming the circuit is half-duplex:
a. How many times must the line turn around in the first six steps of Figure 11–11?
b. How many times must the line turn around in the BISYNC equivalent?

chapter twelve

MULTIPLEXING AND PACKET SWITCHED NETWORKS

In data communications the term *multiplexing* refers to a technique which allows data from a number of low-speed asynchronous terminals to be carried on a single higher-speed circuit. A discussion of multiplexing, as we shall see, naturally leads to the subject of *packet switching,* where we have a complex network of multiplexed circuits rather than a single link. Packet switching technology forms the basis of *public data networks.* In recent years public data networks have been established in virtually every industrialized nation. Internetwork connecting links (called *gateways*) have also been established. These data networks have the ability to carry many different data formats and are therefore not dedicated to equipment manufactured by any particular vendor. These networks have provided data communications devices with a level of convenience and flexibility that the telephone system provided for voice. They have also given computer hobbyists access to sensitive data stored in computer systems operated by large corporations and government agencies.

To understand the reasons why packet switching is such an effective and economical technique, we must understand a little more about the operation of time-shared host computers. The next section is therefore devoted to a brief survey of this subject.

12·1 TIME-SHARED HOST COMPUTERS AND TERMINAL USAGE PATTERNS

Figure 12–1 illustrates a typical time-shared system configuration. We have a number of low-speed asynchronous terminals connected to a fairly large host

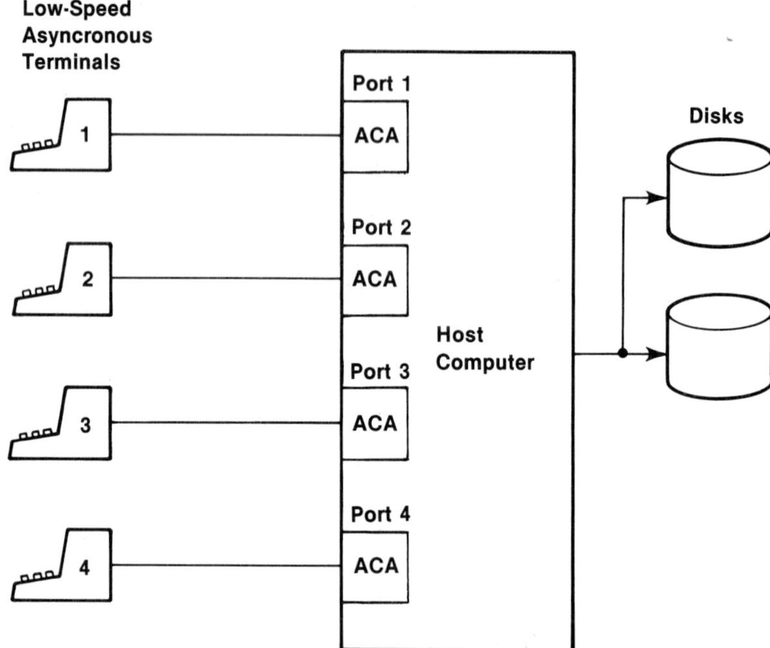

FIGURE 12–1   Typical Time-Shared System

computer via point-to-point links. Only four terminals are shown in Figure 12–1, but moderately sized systems can support fifty or more users. Some of these terminals may be local (directly connected via RS-232C cables), and others may be remote. (We will not clutter the illustrations in this chapter with blocks for modems. Assume that they are in the appropriate places.) In many cases the terminals are used only intermittently, and the remote connections are typically made by dial-up circuits. At the host computer each point-to-point circuit will be supported by a device (similar to the Asynchronous Communication Adapter), which is built around a UART.

From each terminal, a user may initiate and run a program in the host computer. From the user's point of view, all resources of the computer (disks, memory, arithmetic processing) are dedicated to this program. From the host computer's point of view, these services are being provided to many users simultaneously. How is this possible?

In Section 4·9 we described the operational details of a single-user data base system and came to the conclusion that, for many applications involving input from terminals, even a small computer spends most of its time waiting for an external event to occur. In a large system, the effect is even more pronounced. Let us review this process. Figure 12–2 illustrates the host-computer activity associated with a typical transaction. We say that the host computer is "active" when it is executing instructions required to process this particular

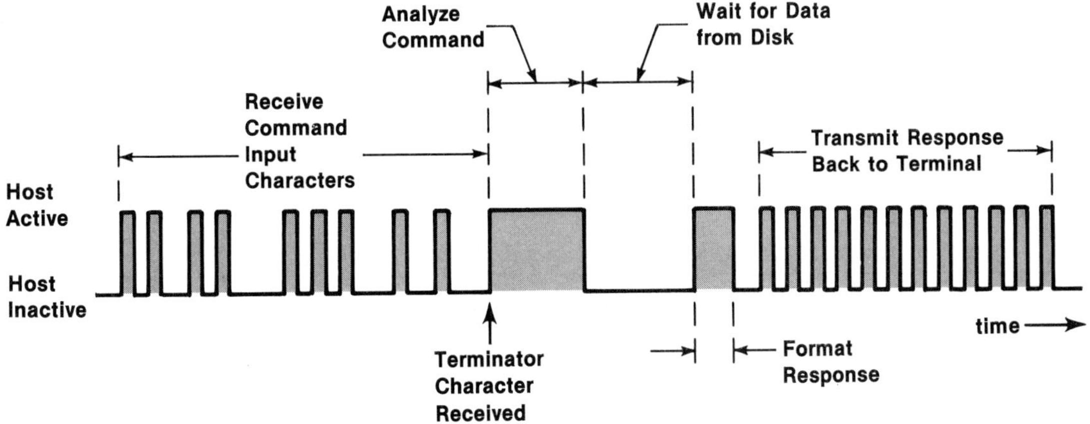

FIGURE 12–2   Host-Computer Activity for a Typical Transaction

transaction. The host computer is considered "inactive" when it is waiting for some event whose time of occurrence is beyond its control.

- The terminal operator types in a command (perhaps a data base inquiry). Each keystroke results in the transmission of one asynchronous character and a receive-data-available interrupt at the host computer. Since the command cannot be interpreted until it has been completely received, the software routine that responds to these interrupts checks each character to see if it is the *terminator* (a specific keystroke, usually carriage return). The terminator indicates the end of the command. If a received character is not the terminator, it is merely stored in memory for later analysis. This process is illustrated at the left side of Figure 12-2. Our terminal operator is not a particularly good typist and the time between input characters is erratic.
- After the terminator has been received, the program can analyze the command and decide how to respond. This will produce a steady burst of host computer activity. Most commands have to do some read or write operations to the disk before they can be completed. The program initiates the disk operation and, once again, must wait until the head is moved to the correct track and the desired data sector rotates into position under the head. At this time, data will finally be moved between the disk and memory.
- When the disk operation is complete, the progam will generally be signaled by another interrupt. The software now formats the response to the original command. This results in another burst of host-computer activity.
- The software initiates transmission of the response to the terminal by writing the first character to the UART Transmit Hold Register. Subsequent characters are sent whenever a transmit-hold-empty interrupt occurs.

Since this transmission is machine controlled, characters will be evenly spaced with minimum stop intervals.

A great deal of idle host-computer time is apparent in Figure 12–2. In fact, if the illustration had been drawn to scale, the percentage of inactive time would appear much greater. If we include the *think time* that the operator spends analyzing the response before beginning the next entry, the idle time is greater still.

This fragmentation of host-computer activity into many intervals, each one of which is measured in microseconds, is the secret of time-sharing. The gaps are filled with work of other user's programs. Time-shared host computers have a large memory. This memory simultaneously holds the programs being run by all "logged in" users (or at least a significant part of each program). The control software of the timeshared host computer (called a *multi-user operating system*) turns control of the host computer over to a user program that has some work to do. While this program is running, some other user programs may become ready to run. They are placed on a waiting list called a *ready queue*. Before long, the currently running program will run out of work, and the operating system will turn the host computer resources over to the program that has spent the longest time on the ready queue. (In actual practice some programs are given higher priority than others.)

This all happens so fast that we get the effect illustrated in Figure 12–3. Each of the four users is running their own *application program* and thinking that they have the host computer all to themselves. The operating system provides a surrounding "environment" and coordinates all application program activites, including access to disks and other peripherals. Of course, if we try to put more and more terminals on this system, the ready queue will grow longer and, at some point, the users will begin to notice things slowing down.

Now that we understand how these systems operate, let us look at the communications costs associated with them.

12·2   LINE COSTS AND MULTIPLEXING

Suppose the host computer in Figure 12–1 is at a corporation's headquarters in some large city. Terminals 1–4 are at a branch office in a distant city. It was not practical to purchase a separate computer for the branch office because users there must have access to the central data base (which may include sales records, accounting data, personnel files) at headquarters. The users therefore dial into the headquarters' host computer several times a week and keep these connections open for hours at a time. Since these calls are placed at times when the highest rates are in effect, the phone bills quickly become astronomical. To make the situation even worse, the average data rate on each

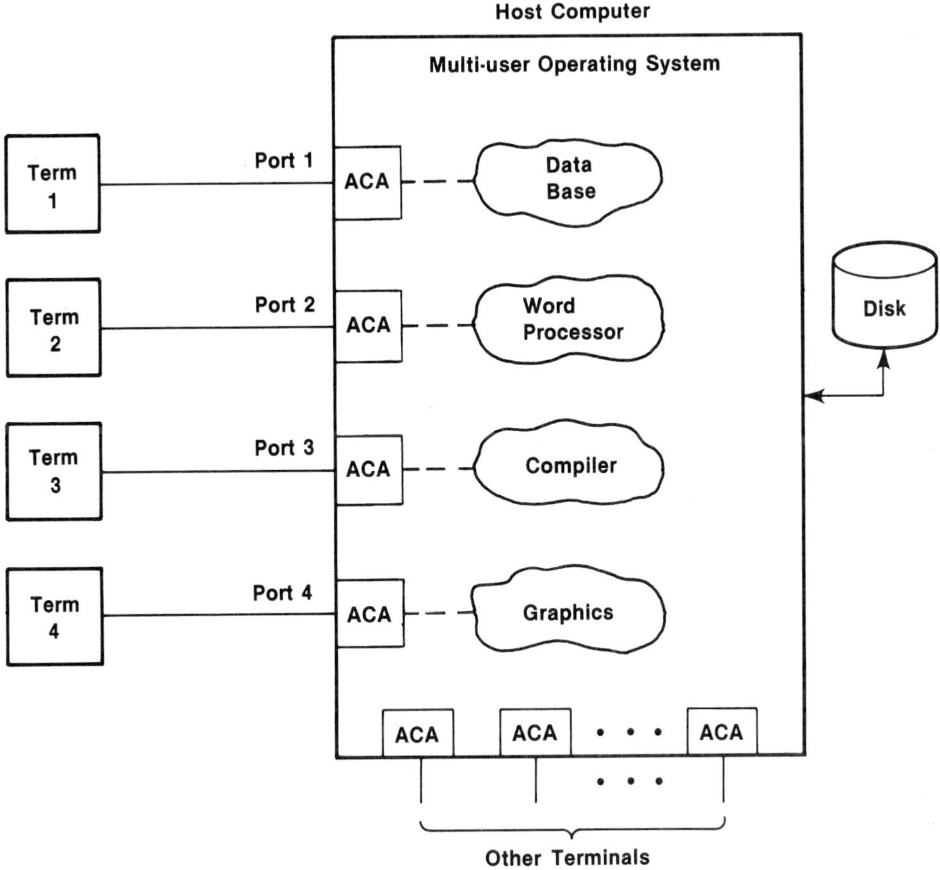

FIGURE 12–3   Time-Sharing

line is much lower than the carrying capacity of the circuit the user's employer is paying for.

How can we take advantage of this available but unused channel capacity? The answer is a device called a multiplexer (the industry standard buzzword is "MUX"), whose function is illustrated in Figure 12–4. We have installed two of these devices: one at the branch office and one near the host computer. The branch office terminals are now locally connected to multiplexer ports, and, at the other end, we have local connections between host-computer ports and multiplexer ports. We have replaced the four dial-up circuits with a single multiplexed trunk. Data transmitted by terminal 1 enters West mux and is sent over the trunk to East mux. East mux routes the data to host-computer port 1. Data from terminal 2 is similarly routed to host-computer port 2, and so on. In the

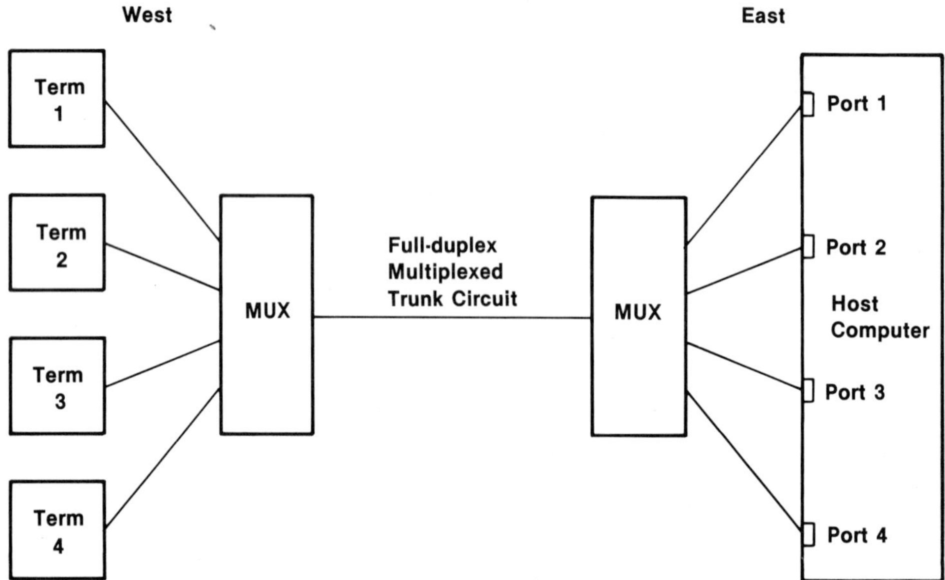

FIGURE 12–4   Communications Multiplexing

opposite direction, data from a host-computer port is routed to its corresponding terminal. Each multiplexer therefore contains both multiplexing and demultiplexing functions (perhaps it should be called a muldem).

Baud rate and data formats on corresponding low-speed circuits are matched at both ends of the multiplexed link, and, from the point of view of the host computer and terminals, the multiplexer is a *transparent* device. All parts of the system operate exactly as they did before the multiplexer was installed. Since all terminals can be operating simultaneously, the multiplexed trunk must be full-duplex, and, since the trunk must be available whenever any terminal operator wishes to contact the host computer, the trunk is usually a dedicated circuit. We have nevertheless replaced four circuits with two multiplexers and one circuit. Since multiplexers are purchased for a one-time cost, and telephone charges are computed on a per call basis, these devices can result in substantial savings.

How do multiplexers work? There are three basic techniques. We will describe them in historical order, earliest first. Figure 12–4 will be the working block diagram for all examples.

12·3   FREQUENCY DIVISION MULTIPLEXING

In *frequency division multiplexing* (FDM) the bandwidth of the multiplexed trunk is divided into a number of channels equal to the number of low-

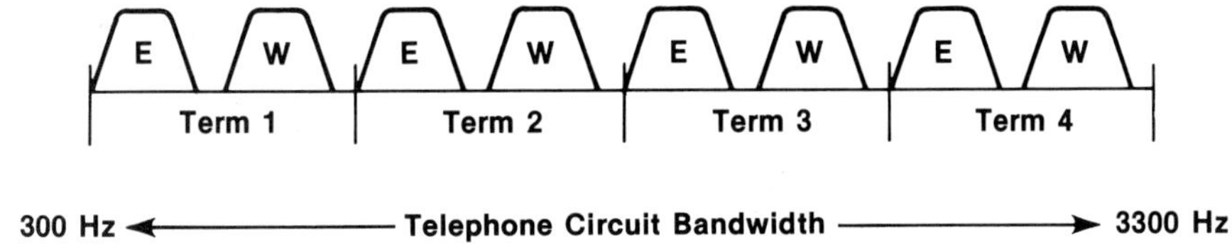

FIGURE 12–5 Frequency Division Multiplexing

speed ports entering the multiplexer. Each of these channels has an eastbound and a westbound carrier, thus providing full-duplex operation (see Figure 12–5). What this amounts to is several modems sharing the same telephone circuit by staying out of each other's way in the frequency domain. Since we have taken the already limited bandwidth of the phone circuit and further subdivided it, the baud rate at which each channel is modulated cannot be very high. Also, the frequency assignments are fixed. As a result, even if a particular channel is idle there is no way for a busy channel to use the idle bandwidth. Frequency division multiplexers therefore handle only a limited number of terminals at low data rates, and they have been generally replaced by devices using more sophisticated techniques.

12·4 TIME DIVISION MULTIPLEXING

The *time division multiplexer* (TDM) accomplishes digitally what the FDM does by analog frequency separation. In this case the multiplexed trunk is used as a single channel driven by high-speed, full-duplex, synchronous modems. The data format on the trunk is shown in Figure 12–6. In each direction we have continuously repeating frames. Each frame starts with a synchronization character or two, followed by a character-sized time slot for each low-speed port entering the multiplexer. If a terminal has produced an outgoing character, the TDM inserts it in the next available slot associated with that terminal. Frames must therefore repeat often enough to avoid loss of data when a terminal or port is transmitting at its maximum character rate. That is, the duration of a frame must be shorter than the duration of a character on any one of the low-speed input ports. If a terminal or port is idle, the TDM leaves its time slot

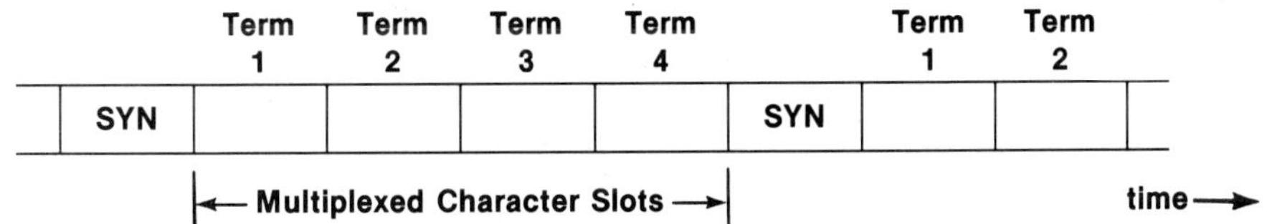

FIGURE 12–6 Time Division Multiplexing

empty. The format of the frame never changes. At the far end the other multiplexer determines a character's destination on the basis of its position in the slot sequence.

TDMs generally provide better performance than FDMs. A single modem makes better use of the telephone channel, because the frequency gaps necessary to separate the numerous FDM carriers do not have to be present. The digital nature of the TDM also makes it more reliable than the FDM, which is primarily analog. TDM time slots do not have to include the stop interval in each character slot, and, since the total number of low-speed lines is usually much larger than the four we have illustrated, the overhead due to the SYN characters is not very noticeable. Nevertheless, if one of the low-speed ports is idle, the TDM has no choice but to let its time slot go by empty. As was the case with the FDM, there is no way that a busy terminal can take advantage of resources not being used by idle devices. Since the observations in Section 12·1 indicate that the all-terminals-busy condition is rare indeed, we must conclude that both TDM and FDM are inefficient techniques. In both cases the multiplexers are designed to accommodate the maximum data rates that can be produced by the attached devices rather than the average data rate, which is much lower. The advent of low-cost microprocessors and memory chips has made possible devices that solve these problems. We will discuss them in the next section.

12·5 STATISTICAL MULTIPLEXING

The *statistical multiplexer* (STAT MUX) is basically a time division device which has the ability to dynamically allocate line time to terminals or ports that are busy while it ignores those that are idle. The device takes advantage of the fact that, on the average, there is lots of idle time on a typical low-speed port (hence the name "statistical"). If you add the baud rates of all terminals attached to the input ports of a statistical multiplexer, the sum will exceed the data capacity of the trunk by a substantial amount. (In a TDM, the trunk capac-

Statistical Multiplexer (16 Port Capacity)
Courtesy of MICOM® Systems, Inc., Simi Valley, CA

ity must be equal to or slightly greater than the sum of the input port baud rates.)

As long as there are idle input ports, this works. What if all ports suddenly became active simultaneously? According to our assumption of low average data rate, the all-busy condition should not last very long, and statistical multiplexers contain memory to "buffer" these occasional overloads. As a result, statistical devices can multiplex many more terminals on a single telephone circuit than either of the older techniques. The only effect an operator may notice is a slight additional delay, and there is no way to determine if this delay is due to multiplexing, host-computer processing, or a combination of the two.

It is obvious that this operation requires precise and sophisticated control. In commercially available statistical multiplexers, this control is provided by a firmware driven microprocessor. Several different design concepts are commonly used. In the following paragraphs we describe just one of these techniques.This is a form of block multiplexing used in packet switched networks. The other techniques deal with the same problems in somewhat different ways.

Figure 12–7 illustrates the transmission activity from four terminals connected to a statistical multiplexer (West in Figure 12–4) and the resultant activity on the multiplexed trunk. The trunk is a full-duplex synchronous channel. Assume that terminals 1 and 3 are character mode devices operated by "hunt and peck" typists, and terminals 2 and 4 are block mode devices which periodically emit a burst of contiguous characters. (All these terminals are asynchronous.) In all cases transmission comes in "logical blocks" ended by a special block termination character. For the character mode devices the termination character

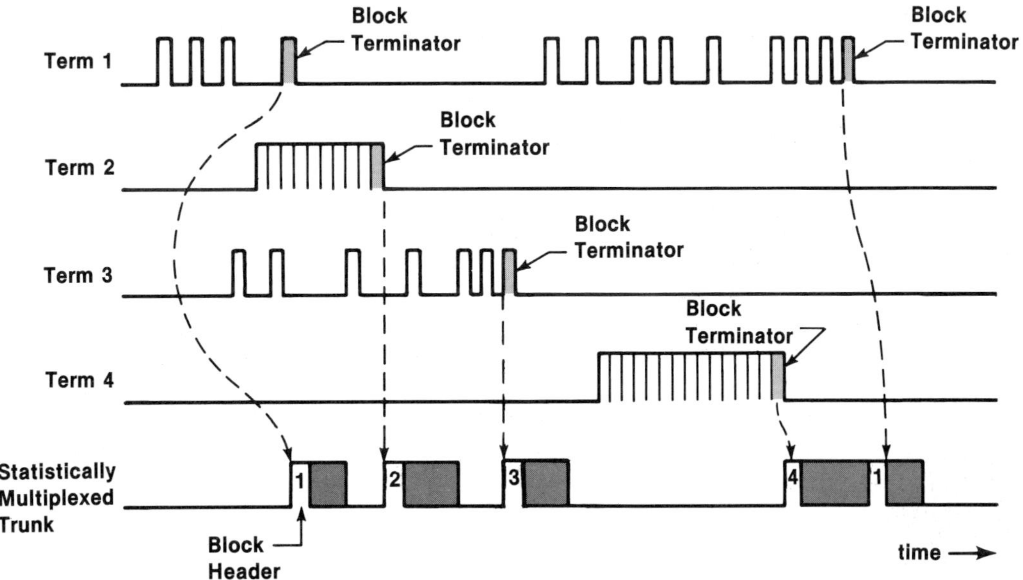

FIGURE 12–7 Statistical Multiplexing

will very likely be "carriage return." In any case nothing is transmitted on the multiplexed trunk until a block termination character is received. (If the character mode devices were depending on a remote echo from the host computer to display typed characters, this echo will not have to come from the multiplexer port.)

The multiplexer accumulates characters from each input port until it detects a block termination character. At this point all accumulated characters for that port will be transmitted on the trunk as a burst. The burst is preceded by a *block header* containing a *terminal identifier* indicating the originator of the block. The multiplexer at the far end examines the block header and, on the basis of the terminal identifier, directs the data to the appropriate port.

This activity is illustrated in Figure 12–7. Each time a terminal sends a block terminator, a block is transmitted on the trunk. If the trunk is idle, the block is transmitted immediately. If the trunk is busy (as is the case at the time of the second block terminator sent by terminal 1), the block is transmitted as soon as the trunk becomes free.

Statistical multiplexers have a degree of flexibility and programmability not present in the earlier multiplexers. The user can typically select, on a port-by-port basis such parameters as

- baud rate
- data format
- echo/no echo
- block termination character (blocks may also be dispatched if the accumulated character count exceeds some operator-selected value, even if a termination character has not been received)

12·6 "SOFTWARE MULTIPLEXERS"

In all of the previous examples Figure 12–4 was the common block diagram. This configuration is convenient for the situation where a multiplexer is installed as a replacement for existing individual connections. However, does it really make sense to fan out the low-speed lines to individual ports at the host computer end? From Figure 12–3 we can draw the following conclusions:

- The association of a terminal with a host-computer port is really equivalent to an association between a terminal and an application program.
- Our host computer is capable of running many programs simultaneously.

In view of this information we can build the configuration illustrated in Figure 12–8. We have directly connected the statistically multiplexed trunk to a host computer port (in this case the port must be synchronous), and the hardware multiplexer has been replaced by a software multiplexer. The software multiplexer is just another program that runs in the time-shared host computer. It examines block headers and, on the basis of the terminal identifier,

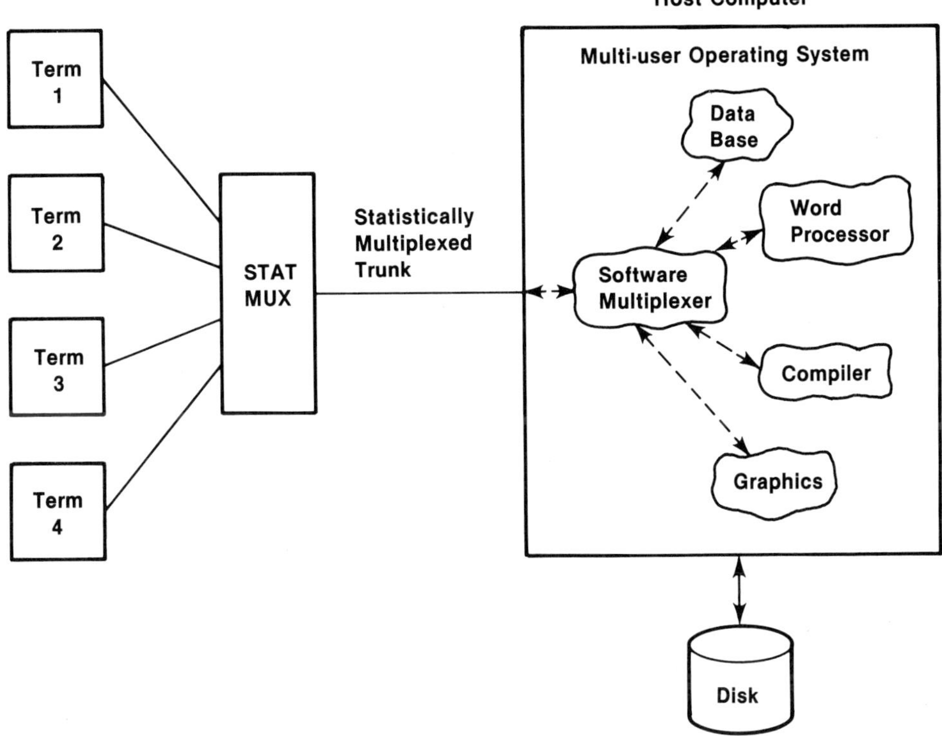

FIGURE 12–8 Eliminating a Multiplexer

delivers data blocks to the appropriate applications program. It also accepts data from applications programs, places a block header in front of them, and transmits them on the multiplexed trunk. These transfers between applications programs and the multiplexer are now accomplished by moving data around in the host-computer memory rather than by moving serial characters over an RS-232C cable.

In Figure 12–8 we now have all the concepts necessary to study packet switched, public data networks. Network technology has its own unique vocabulary, so we will have to redefine some of the terms we have been using:

- The data blocks on the statistically multiplexed trunk will be called *packets*.
- The terminal identifier in the block header (packet header) will be called a *logical channel number* (LCN).
- The statistical multiplexer will be called a *packet assembler/disassembler* (PAD).
- We previously stated that the presence of a multiplexer was transparent to a terminal operator. That is, from the operator's point of view a direct connection between the terminal and the host computer still appears to exist. This apparent connection will be called a *virtual circuit*.

12·7   PACKET SWITCHED NETWORKS

Figure 12–8 does not look like much of a network, but all the basic elements are there. In Chapter 1 we defined a network as a system which provides its attached users with the ability to establish communications with other users. In Figure 12–8 the "users" are terminals and applications programs. From this point of view, the statistical multiplexer in Figure 12–8 allows the establishment of virtual circuits between users. This is still a very limited network, but it takes only a small leap of the imagination to expand this structure to the configuration shown in Figure 12–9.

In Figure 12–9 we have a rather large collection of host computers and PADs. Rather than directly connecting any two such devices as we did in Figure 12–8, we have connected each PAD or host computer to a *packet switching node*. The statistically multiplexed circuit, which now crosses the boundary between the outside world and the network, has become a *network access link*. The packet switching node is a device (usually a computer) which has the ability to establish a virtual circuit between any pair of users. This is usually a connection between a terminal and an applications program, but there is no reason why a virtual circuit cannot be set up between two applications programs or even between two terminals.

In many respects there are analogies between the data network in Figure 12–9 and the voice telephone network we discussed in Chapter 7. The packet switching nodes are equivalent to end offices; the applications programs and terminals are subscribers. There is no direct analogue to the subscriber loop, because a subscriber loop is usually dedicated to a single user. The network access link is more like a party line. To further extend this metaphor, virtual circuits come in two varieties:

- *Switched virtual circuits* (SVCs) are temporary connections between users. They are established at the request of a user who must specify the *network address* (telephone number) of the terminal or program the user would like to contact. As is the case with switched telephone calls, SVCs may be disconnected by the user at either end of the call. Users are billed for SVC connections on the basis of connect time and the number of packets transmitted. Typical telephone features, such as reverse charging, are available.
- *Permanent virtual circuits* (PVCs) are long-term connections analogous to dedicated circuits in the telephone network. The cost of a PVC is usually a fixed monthly rate.

The protocols, packet header formats, and virtual circuit setup procedures for the network access link have been standardized in accordance with a specification called X.25. This document was published by the CCITT, the international organization we discussed briefly in Chapter 8. Public data networks

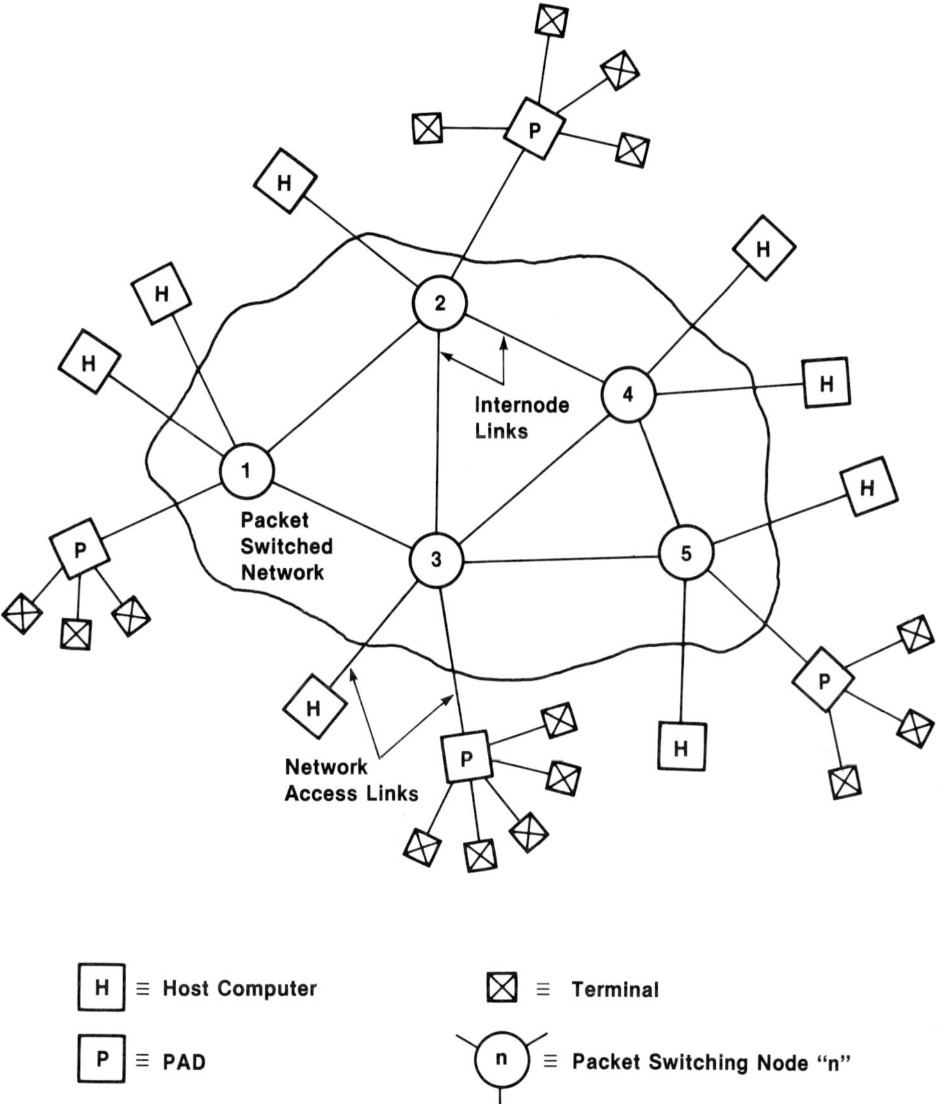

FIGURE 12–9 Packet Switched Network

based on packet switching technology have been established in many coun-
tries, and X.25 has been almost universally adopted as the standard for interface
to these networks. Numerous manufacturers are building X.25 compatible
communications equipment. We will have more to say about the details of X.25
later in this chapter.

What goes on inside the packet switched network, and what is the nature of the links between the nodes? From the user's point of view these questions are irrelevant in the same sense that a telephone subscriber is not concerned about the phone company's decision to carry the subscriber's voice on a microwave link or a fiber-optic cable. It is also likely that internal operation will vary from one network to another. As long as the interface complies with X.25, devices compatible with this specification will be able to use the network. We will, however, make some general statements about the internal operation of these networks.

In most cases the internode links are statistically multiplexed, full-duplex, synchronous circuits using a packet format that is almost identical to the format used on the network access link. The key to maintaining virtual circuits on a packet switched network is the logical channel number (LCN). Each link (network access or internode) is assigned a fixed number of LCNs. On internode links, the LCNs are roughly equivalent to interoffice trunks in the telephone system. A virtual circuit is merely a list of LCNs, one for each link the connection must traverse. For example, suppose we wanted a virtual circuit between the pad on node 5 and the host computer on node 2. The virtual circuit could be specified as follows:

LINK	LCN
Pad to Node 5	7
Node 5 to Node 4	12
Node 4 to Node 2	9
Node 2 to Host	3

The selection of LCNs is pretty much arbitrary, as long as the one selected is free at the time the virtual circuit is established. The nodes contain *routing tables* which store the associations between LCNs on their various links. For example, the entry in the node 4 routing table for the virtual circuit described earlier might look like this:

Link	LCN	↔	Link	LCN
Node 5 to Node 4	12	↔	Node 4 to Node 2	9

This is one entry in a table of many such entries. It indicates that the link/LCN pair on the left is "connected" to the link/LCN pair on the right. This means that nodes must change the value of the LCN in packet headers before they retransmit them toward their destination. The routing table entry instructs node 4 to do the following when a packet with LCN = 12 arrives from node 5:

- change the LCN value in the packet header from 12 to 9
- transmit the packet on the link to node 2

In the opposite direction, the LCN conversion would be from 9 to 12. The nodes perform this routing function simultaneously for many virtual circuits. In a similar manner PADs and software multiplexers must keep track of associations between LCNs and users.

These routes are determined at the time the virtual circuit is established. For PVCs they are permanently allocated by the people who operate the node. For SVCs, they are dynamically assigned by node software. In the previous example suppose the request for a virtual circuit was initiated by the PAD connected to node 5. At this point, node 5 could have chosen to route the virtual circuit by way of node 3 rather than node 4. The decision may have been based on the availabilty of LCNs or the instantaneous traffic load on the internode links.

12·8   DATA FORMAT ON AN X.25 LINK

In Figure 12–7 we illustrated data blocks on a statistically multiplexed trunk, and we later stated that these blocks were equivalent to packets. This is a true, but oversimplified, picture of the actual appearance of packets on an X.25 link. There are a number of basic communications issues that we have thus far ignored in our discussion of packet transmission. These are

- framing
- error control
- line control
- flow control
- sequence control

But we already know how to deal with these problems. They are exactly the features of a protocol that were described at the beginning of Chapter 11. The protocol selected by the CCITT for X.25 is HDLC. All of these issues are neatly handled by placing the packet inside an "envelope." The envelope, of course, is the HDLC frame structure (see Figure 12–10A). In this case the HDLC class of procedure is called *asynchronous balanced mode* (the one we discussed in Chapter 11 was "normal response mode"). Do not be confused by the "asynchronous" in the name of the class of procedure. The data format of an X.25 link is definitely synchronous. The term refers to the fact that this class of procedure allows any station on the link to transmit a frame at any time. This is exactly what we want for the point-to-point, full-duplex links used for X.25 operation. Otherwise, the HDLC frame types and procedures are identical to those described in Chapter 11.

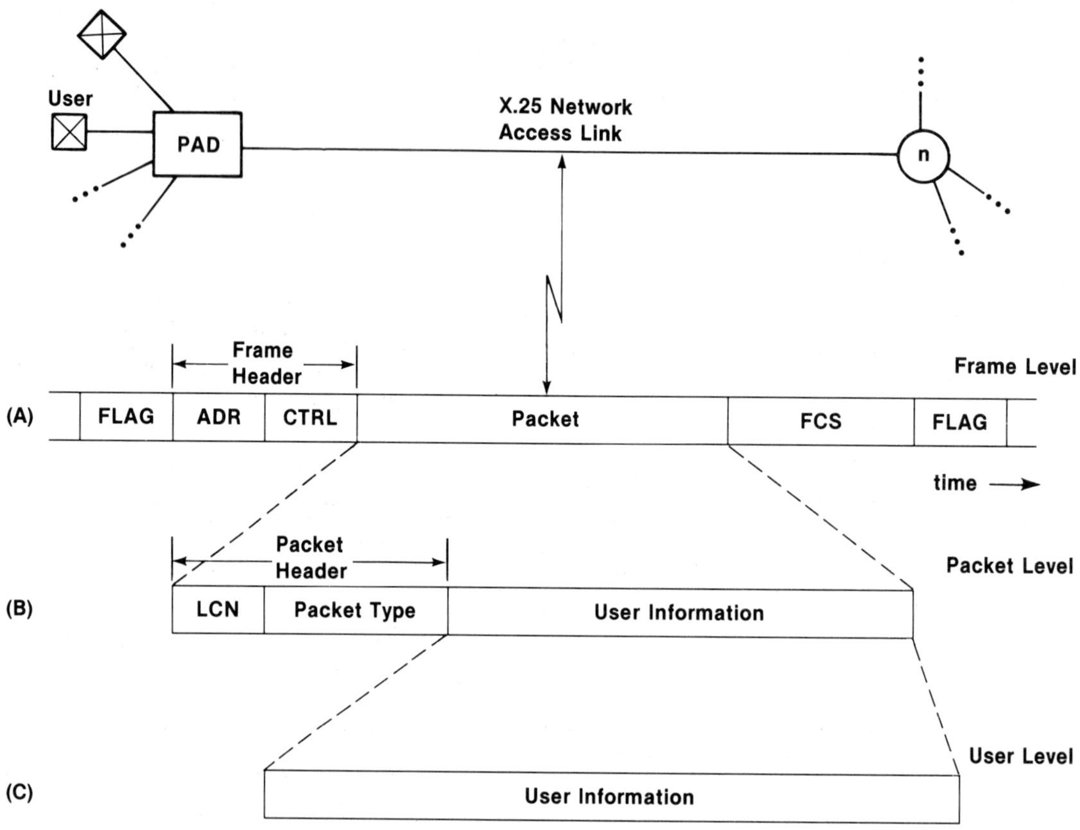

FIGURE 12–10   Data on an X.25 Circuit

The packet can be extracted from the information field of the HDLC frame (see Figure 12–10B). The packet contains a *packet header* and *user information*. The packet header contains

- the logical channel number (LCN)
- a packet type identifier (thus far we have only discussed data packets; there are a few additional types that we will mention later in this chapter).

The user information is the actual data generated by a terminal or application program for transmission to another user at the far end of the virtual circuit. The user information (see Figure 12–10C) is embedded in the packet in a manner similar to the way the packet was embedded in the HDLC frame. Because of this "nesting" it is customary to view the X.25 interface as a succession of layers. This concept is illustrated in Figure 12–11.

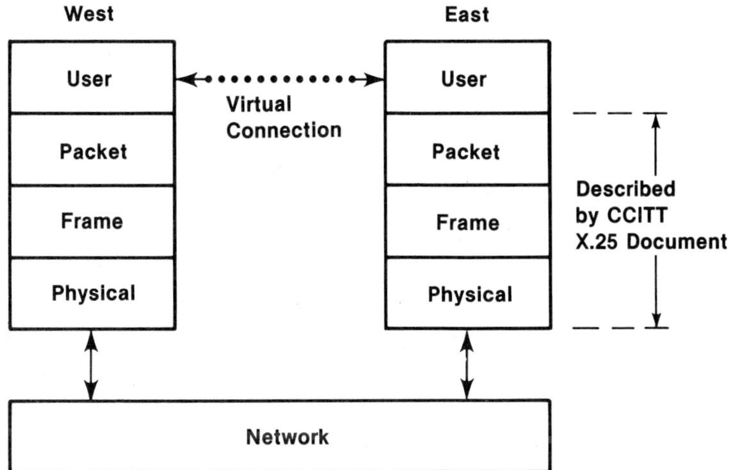

FIGURE 12–11   The Layered Structure of X.25 Network Interface

At the West end of the network, a user generates a "logical message" which is passed down to the West packet layer. The packet layer will

- break the logical message into a number of packet-sized blocks;
- build the packet headers and insert the correct LCN;
- pass the packet down to West frame layer.

The frame layer is unconcerned with such things as LCNs and packet type. It simply inserts each packet into an HDLC information transfer frame with the appropriate NS/NR values and sends it to West physical layer. The physical layer is merely the RS-232C interface to the network access line.

At the other end of the network, the packet is received by East physical layer and passed to East frame layer. The frame layer

- sends any acknowledgments required by the HDLC protocol;
- extracts the packet from its envelope and passes it up to East packet layer.

East packet layer

- strips off the packet headers;
- assembles the packets into a logical message;
- passes the logical message to the user indicated by the LCN in the received packets.

To summarize, a logical message generated by one user is delivered, with unchanged format and content, to another user. Neither user had to be concerned with the details of packetization, LCNs, HDLC or any other communications procedures. The layered "architecture" of X.25 has thus provided a "virtual connection" between East user and West user.

The CCITT X.25 document actually describes network interface procedures in terms of physical, frame, and packet layers. The internal components (hardware or software) of PADs and other X.25 compatible devices can be divided into distinct sections associated with the functions of these layers.

12·9   THE PACKET LAYER

The packet header is always 3 bytes long. It contains the LCN, a packet type identifier, and a few miscellaneous fields that we will not bother describing. The packet types fall into a few different categories:

DATA   These packets carry the user information we have been discussing. They are usually limited to some maximum number of bytes by network rules. The headers of these packets contain a "more" bit, which is turned on to indicate that there are additional packets (for the current LCN) which must be combined to form a logical message.

FLOW CONTROL   The frame level HDLC protocol deals with flow control on the X.25 network access link. The packet layer provides a method of end-to-end flow control for an individual virtual connection (the user terminal may be a printer). This involves RECEIVE READY and RECEIVE NOT READY packets which are used in much the same manner as their counterparts at the frame level. NS and NR counts are maintained for each virtual connection by the packet layer software.

CALL SETUP AND CALL CLEARING   There are several different packet types associated with the setup and disconnection ("clearing") of switched virtual circuits (SVCs). We will see how some of them are used in the next few paragraphs.

When a new X.25 network access link is implemented between a subscriber and a node, a fixed number of LCNs is assigned to this link. The LCNs are numbered, starting at 1. If any permanent virtual circuits (PVCs) are present, they are assigned to a contiguous block of LCNs at the bottom of the range. The remaining LCNs form a pool. LCNs from the pool are assigned to SVCs as connections are established and returned to the pool when connections are cleared.

A brief history of a virtual call is illustrated in Figure 12–12. The sequence starts when a terminal operator sends a special message to the PAD indicating the desire to initiate a virtual call to a particular host computer. The operator must also provide the network address of the host computer the operator wishes to contact. If a free LCN is available, the PAD issues a *call request* packet

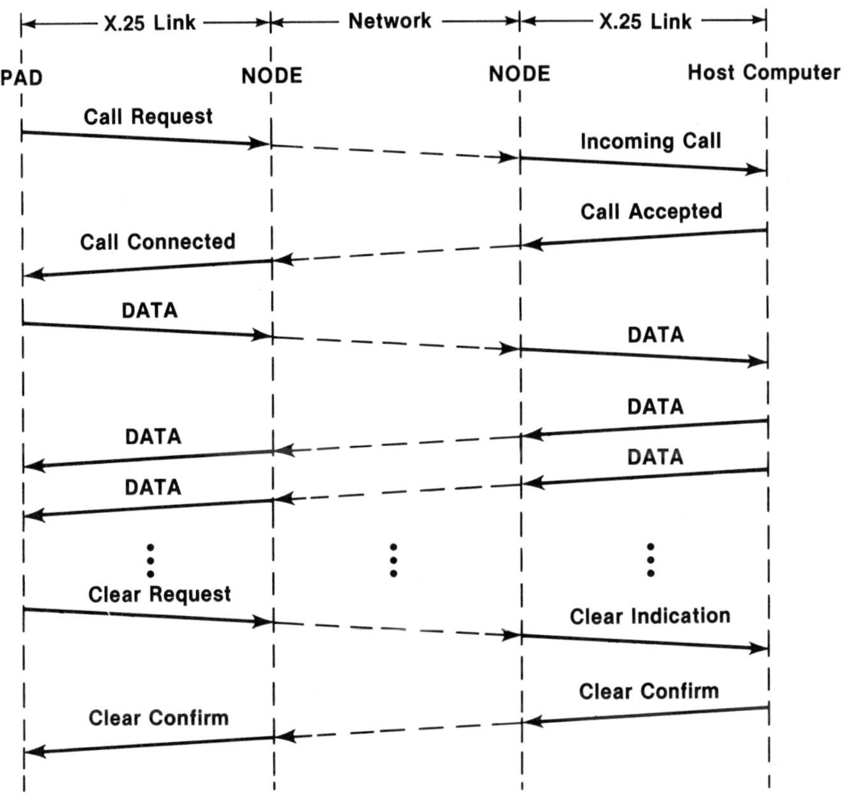

FIGURE 12–12   The History of a Virtual Call

on this LCN. If the connection is completed, this LCN will be associated with the PAD end of the call until it is cleared. The information field of the call request packet contains the network address supplied by the terminal operator.

The network checks to see if there is a free LCN on the X.25 link to the addressed host computer. If there is, the network will issue an *incoming call* packet on this LCN. The host computer has the option of accepting or rejecting the call. In our illustration the host computer issues a call accepted packet on the same LCN used for the incoming call packet. This LCN will be associated with the host-computer end of the connection for the duration of the call. The network translates the call accepted packet into a call connected packet to the PAD. The SVC is now complete.

Data packets can now be exchanged in both directions. At some point the terminal operator decides to end the session and sends another special message to the PAD indicating a desire to disconnect. In response, the PAD issues a *clear request* packet to the network. Since all links of the virtual circuit are now identified by LCNs in the routing tables of various nodes, the clear request packet does not have to carry any address information. The network translates the clear request (which carries the LCN associated with the PAD end) into a *clear indication* packet (which will carry the LCN associated with the host-computer end of the call). The host computer will perform whatever internal processing is required to disconnect the application program from the virtual circuit and issue a *clear confirm* packet. At this point the LCN for the host-computer end is returned to the free pool. The network sends the clear confirm across to the PAD, and the LCN at the PAD end of the call will also be freed.

REVIEW QUESTIONS

1. Eight asynchronous terminals are attached to a TDM's input ports. Four operate at 300 baud and four at 110 baud. All use the same data format: 7 data bits, odd parity, 1-bit stop interval.
    a. At what rate must the TDM frames repeat to guarantee that no data will be lost from any input port?
    b. Assuming that theTDM frame consists of two 8-bit SYN characters and a 9-bit (start, data, parity) slot for each input port, how many bits are in the TDM frame?
    c. What is the minimum baud rate at which the multiplexed trunk can be operated?
    d. Given the previous parameters for frame rate and multiplexed trunk speed, would it make any difference if all eight ports operated at 300 baud?
    e. Why is it necessary to carry the start bit in the character slots of the TDM frame?

2. A statistical multiplexer has sixteen input ports. These are connected to identical 1200-baud host-computer ports. The data format for the host-computer ports is 7 data bits, odd parity, 1-bit stop interval.   The multiplexed trunk operates at 4800 baud.
    a. If all ports become active simultaneously at the maximum character rate, what is the difference (in bits per second) between the total rate at which data is entering the multiplexer and the rate at which data is leaving on the multiplexed trunk? (Assume that the multiplexer must transmit all bits that come in except stop intervals, and ignore the "overhead" due to block headers.)
    b. If this all-busy condition persists for 1 second, how many bytes of memory does the multiplexer need to buffer the overload?

3. In the packet network illustrated in Figure 12–9 we have the following virtual circuits connected:
   I. the PAD on node 1 to a host computer on node 4
   II. the PAD on node 3 to the host computer on node 2
   III. the PAD on node 5 to the host computer on node 2
   IV. the host computer on node 3 to a host computer on node 1
   a. Assign internode links and LCNs to these connections. Assume that all LCNs are initially free and LCN assigments for each internode link are in a contiguous group starting with 1.
   b. Based on these LCN assignments, specify the routing table entries for nodes 2 and 3.

4. Suppose a network access link to a PAD was given fewer LCNs than the number of input ports to the PAD.
   a. What are the possible effects of this on the operator of a terminal attached to the PAD?
   b. Compare this with the telephone network congestion problem described in Chapter 7.
   c. Would it make sense to connect a PAD to a network access link with more LCNs than the number of input ports? If not, why?

5. Which X.25 layer would deal with the following situations:
   a. assignment of a free LCN to a new call request
   b. an HDLC frame check sequence error
   c. accepting or rejecting incoming call packets
   d. generating correct voltage levels on the interface to the network access link modem
   e. HDLC zero stuffing
   f. clearing the call on an SVC
   g. detecting loss of carrier on the network access link

chapter thirteen

LOCAL AREA NETWORKS

13·1   THE NEED FOR LOCAL NETWORKING

In recent years organizations such as corporations, universities, and government agencies have acquired large collections of single-user-oriented digital hardware (personal computers, word processors, etc.). It soon became clear that isolated data processing often left some things to be desired:

- Users wanted the ability to send messages to any other user in the organization.
- It did not make economic sense to dedicate an expensive device, such as a high-quality graphic printer, to a personal computer, but many users had occasional need for such devices.
- It is extremely difficult to distribute multiple copies of a rapidly changing data base, such as the employee records of a large corporation. It makes more sense to keep the data base at one central location. This assures that all users instantly see the most current version of the data.
- Users occasionally wanted access to a large host computer in order to run programs that were too big to be handled by a personal computer.

One possible solution is to provide each personal computer with terminal emulation programs and connect all users to a single host computer via RS-232C cables and/or modems. In many organizations, however, the total number of users is too large for a single host computer. Also, because organizations grow in a haphazard manner, there are usually a number of existing host computers with different functions. An engineering manager, for example, may need to access personnel records and also the ability to run complex mathematical calculations. Secondly, the enormous amount of cabling required for a system of this type is expensive to install and difficult to rearrange.

As an alternative, the organization could choose to operate a packet switching network similar to those described in the last chapter. This provides all users with unlimited access to other users and devices, but packet switching nodes are expensive, complex devices which usually must be operated by experts.

What the organization typically needs is a network with the following properties:

- dynamically established communications links among several hundred users;
- inexpensive installation, high reliability, no need for expert operators;
- simple reconfiguration procedures for addition of new users and relocation of existing ones.

An excellent choice for this application is a broadcast network (as described in Chapter 1). In such a network, any user has access to all others without the requirement for switching nodes or a centralized controller. New users may be attached by simply inserting a new "antenna" into the medium. The obvious question is, what shall be used as a medium? As long as all users are located within a radius of a mile or so (this is a *local area* network), a coaxial cable is an excellent choice:

- The bandwidth of coaxial cable runs from DC to several hundred megahertz.
- There is virtually no leakage of signal from the coaxial cable and virtually no pickup of external interference. (Any airborne radio frequency broadcast must comply with strict regulations concerning frequency and power level.)
- Users can be added and moved around by installing taps at any point along the cable. (Actual systems impose some limitations on the number of users and where they can be attached. We shall have more to say about this later.)

As a result of these obvious advantages, a number of local area network (LAN) concepts have been developed with coaxial cable as the transmission medium. Some of these are *baseband systems* (which place digital signals directly on the coaxial cable), and others are *broadband systems* (which use modulated carriers). Conceptually, however, these systems have many similarities. Rather than survey several of these implementations, we will select one and study it in some depth. The one we study is called *Ethernet*.

13·2 ETHERNET HARDWARE

The specification for Ethernet was jointly published in 1980 by Digital Equipment Corporation, Intel Corporation, and Xerox. It is a baseband system, but, for historical reasons, signals on the coaxial cable are referred to as "carrier." The data rate on the cable is 10 Mbit/s, which is orders of magnitude higher than anything we have previously encountered.

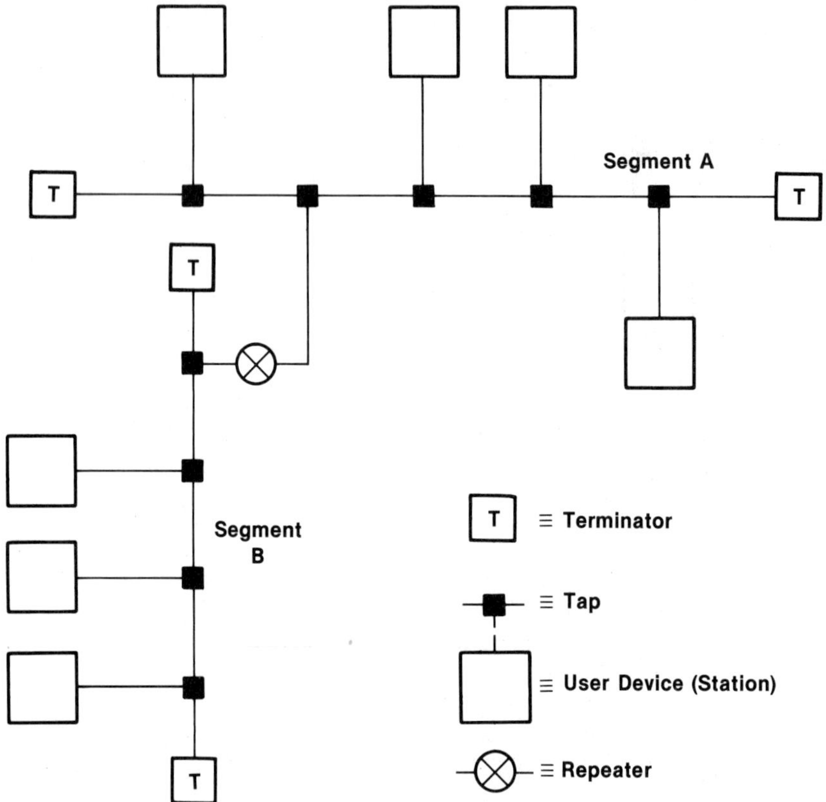

FIGURE 13–1   A Local Area Network Based on Coaxial Cable

A typical Ethernet system is illustrated in Figure 13–1. We have some lengths of coaxial cable with terminators at each end. These lengths are called *segments*. To prevent reflections and other signal distortions, the terminator must be a resistor whose value, in ohms, is equal to the characteristic imped-ance of the coaxial medium (see Section 8·4). At points along the cable, user devices (*stations*) are attached by *taps*. When a station transmits, its signal propagates in both directions from its tap point until it is absorbed by the ter-minators. In order to increase the total length and number of users on the net-work, segments may be interconnected by *repeaters*. A repeater appears as a tap in each of the connected segments and has the ability to amplify signals from either segment.

There are numerous rules related to the allowable configurations of Eth-ernet systems. To present an idea of the limitations of LANs, we list some of the more significant rules:

- The maximum length of any segment is 500 m.
- The outer jacket of the coaxial cable is marked at regular 2.5-m intervals. Taps may be attached only at these marks. (Closer spacing results in disruption of the cable's characteristic impedance.)
- The path between any two taps may not include more than two repeaters.

The last rule has some interesting implications. In Figure 13–2 the path between any station on segment A and another station on segment C includes two repeaters. We therefore cannot attach additional repeaters between segment A and a fourth segment since this would introduce three repeaters between a station on the new segment and any station on segment C. In fact, the only segment in Figure 13–2 on which additional repeaters can be introduced is segment B. If we disregard the length of the cables to and from the repeaters, the two-repeater rule, in combination with the 500-m segment rule, limits the maximum separation between taps to 1500 m. (The use of a special repeater can increase this maximum separation to 2500 m.)

Figure 13–3 takes a closer look at the hardware between the tap and the user station. In actuality, the tap is combined with a few electronic circuits called the *transceiver*. The user station is connected to the transceiver/tap by a transceiver cable whose length cannot exceed 50 m. The cable contains four twisted pairs:

- Transmit: data from the user to the network
- Receive: data from the network to the user
- Collision presence: a signal which indicates that two devices are simultaneously transmitting data to the network (we describe the use of this signal in Section 13·4)
- Transceiver power: The user station must provide DC power to the circuits in the transceiver

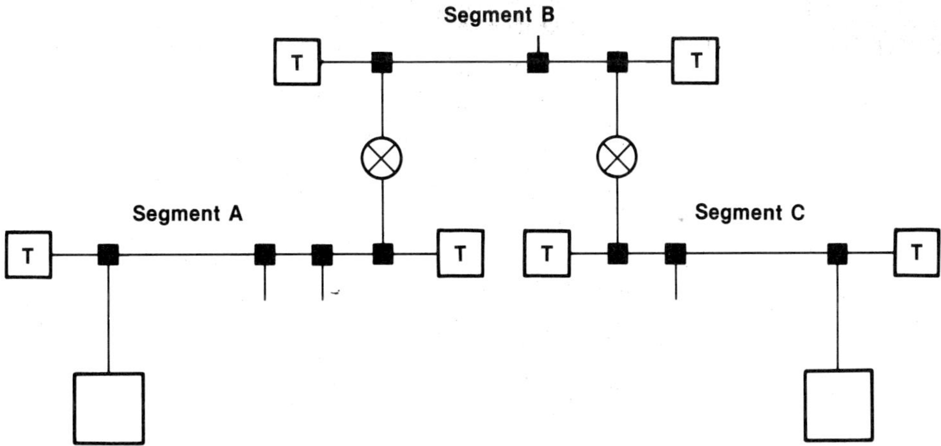

FIGURE 13–2   Multisegment Ethernet

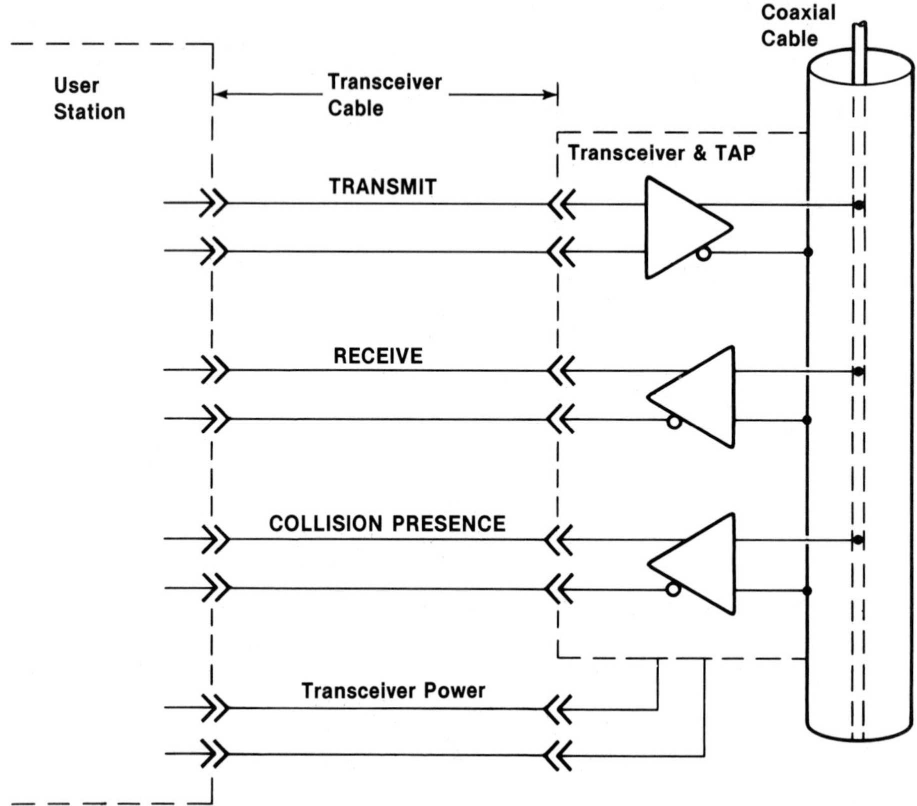

FIGURE 13–3   Connection from Tap to User Device

The interface devices for these signals are balanced drivers and receivers. Considerations of cable length and speed discussed in Chapter 8 would indicate that for a 50-m cable operating at 10 MHz, there is no other choice. The data on the transmit and receive pairs (and also on the coaxial cable) is in a format called *Manchester encoding*. A Manchester encoded waveform is illustrated in Figure 13–4. Time is divided into equal bit intervals (100 ns for a 10-MHz bit rate). At the center of each interval, a state transition always takes place. The direction of the transition indicates the polarity of the data bit represented by each interval:

$$1 = \text{positive-going transition}$$
$$0 = \text{negative-going transition}$$

At the end of the interval, the signal goes to whatever state is required to set up the transition that will occur at the next midinterval point. There are conceptual

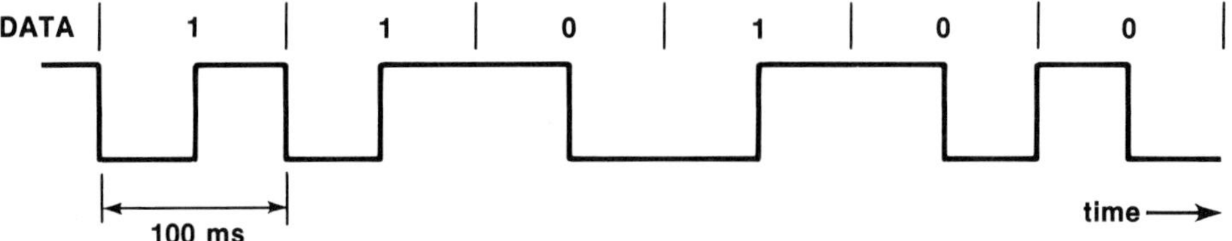

FIGURE 13–4 Manchester Encoded Data

similarities between Manchester encoding and the modulation techniques for synchronous data described in Section 9·2:

- the division of time into equally spaced bit intervals
- the guaranteed occurrence of an event (the positive- or negative-going transition) during each bit interval

As a result Manchester encoded waveforms, like DPSK modulated signals, carry both timing (clock) and data elements.

13·3 ETHERNET DATA FORMAT

As is the case in some of the other protocols we have seen, data on Ethernet comes in blocks called frames. Like X.25, Ethernet is a layered protocol, and data blocks generated by higher layers are surrounded by headers and trailers produced by the lower layers. The layers of Ethernet are shown in Figure 13–5.

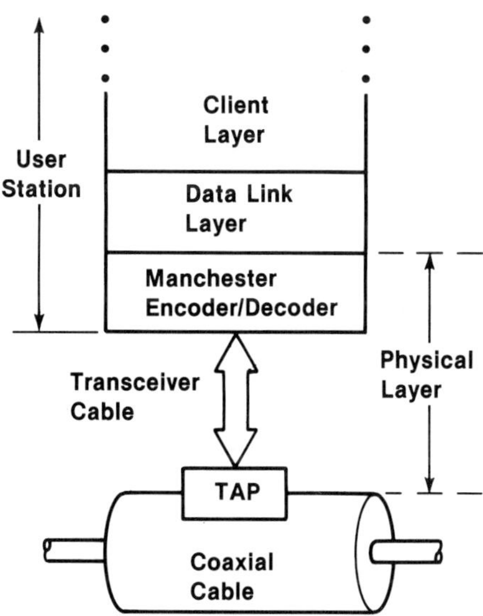

FIGURE 13–5 The Layered Structure of Ethernet

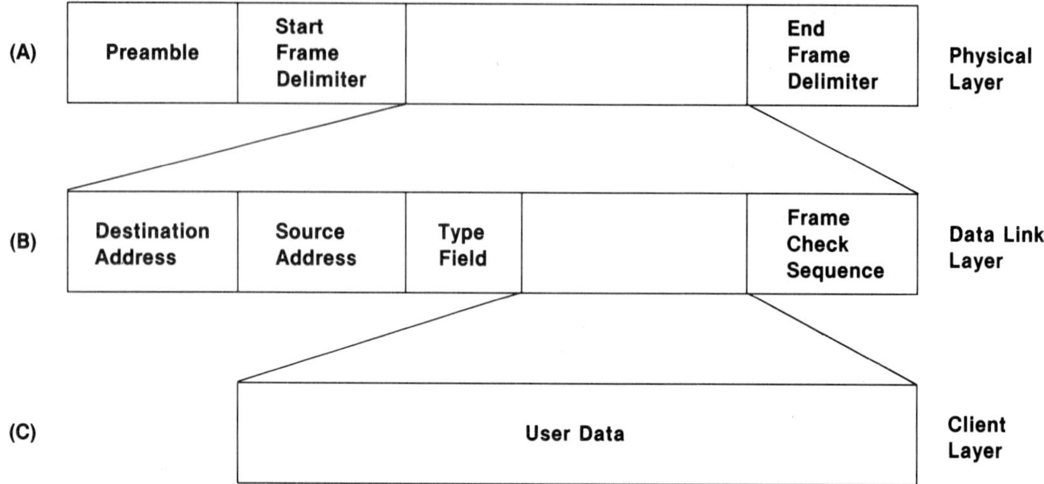

FIGURE 13–6   Ethernet Frame Structure

- The *client layer* represents the user and is not described in the Ethernet specification.
- The *data link layer* handles station addressing and error detection and operates under a protocol described in Section 13·5. This layer is usually constructed from a combination of hardware and software components and resides totally in the user station.
- The *physical layer* function is roughly equivalent to that of a modem in telephone data communications. It generates signal levels appropriate for the medium, and extracts data and synchronization from the signals on the coaxial cable. There is no software in the physical layer. It includes
    a. the transceiver and tap
    b. the transceiver cable
    c. the Manchester encoder/decoder (which usually resides in the user station)

Just as X.25 created virtual connections between users above the packet layer, Ethernet allows client layers to communicate without having to deal with the details of the network.

The structure of the Ethernet frame and the relation of its components to the protocol layers are illustrated in Figure 13–6.

Physical Layer Components

The *preamble* is a 62-bit sequence of alternating 1s and 0s. This allows receiver circuits to synchronize with the Manchester encoded waveform. This is immediately followed by the *start frame delimiter,* which is merely two consecutive 1 bits. The *end frame delimiter* is really not a data pattern. Since Manchester encoding produces steady activity (signal state transitions) while the frame is in progress, a period greater than 1 bit interval without any activity is indicative of the end of the frame.

Data Link Layer Components

Since frames transmitted on a broadcast medium are received by all attached devices, source and destination addresses are necessary to identify the ends of a "virtual connection." In the Ethernet frame each of these fields is 6 bytes long. Each station is uniquely identified by a physical address. The protocol also allows *multicast* (group) addresses to be defined, and one pattern may be defined as a broadcast (all stations) address. When a station receives a frame whose destination address field contains the physical address of a different station or the multicast address of a group to which it does not belong, it must ignore that frame.

The 2-byte *type field* is unused by the data link layer. It is placed in the frame for the convenience of the client layer. The *frame check sequence* is a 32-bit CRC. The generator polynomial is

$$X^{32} + X^{26} + X^{22} + X^{16} + X^{12} + X^{11} + X^{10}$$
$$+ X^{8} + X^{7} + X^{5} + X^{4} + X^{2} + X + 1$$

(The questions at the end of this chapter will not include any manual BCC calculations based on this polynomial.)

The user data field is the part of the frame generated by and delivered to the client layer. In Ethernet, this field must be at least 46 bytes, and no more than 1500 bytes, long. The 46-byte minimum for user data forces the total data-link-layer frame size to be at least 64 bytes long. The significance of this will be explained in the next section.

13·4 ETHERNET PROTOCOL

In Chapter 11 we described the line control function of a protocol as a set of rules which prevent devices that share a medium from interfering with each other's transmissions. Since the Ethernet coaxial cable is half-duplex, this means that only one station may transmit at any given time. On the broadcast network, this must be accomplished without polling by a master station.

The line control protocol used by Ethernet is called Carrier Sense Multiple Access with Collision Detection (CSMA/CD). It operates in a manner roughly analogous to the way people use a citizen's band radio channel:

- A station with the frame to send may initiate transmission any time that it does not detect the presence of another station's carrier on the cable.
- Because of propagation delays on the cable, a station may not become aware of another station's carrier until some time after it has started its own transmission. This is a collision. When a collision is detected (this is the purpose of the *collision presence* signal described in Section 13·2), both stations stop their frame transmission.
- The stations then back off for a period of time and try to retransmit. If both stations waited approximately the same time, it is likely they would collide again. The delay algorithm used by the stations is therefore based on a random number generator. This makes a second collision (between the same pair of stations) unlikely.

It is obvious that all collisions must be detected. This is the reason for the minimum frame size. It can be demonstrated that if all frames last at least twice the time required for a signal to propagate from end to end in the largest possible network, all collisions will be detected. A simplified discussion of why this is true follows.

Figure 13–7 shows a single 500-m cable segment with three user stations. For simplicity assume that the distance between stations A and B is negligible. Also assume that A and B are very close to the left terminator and C is so close to the right terminator that we can say C is exactly 500 m from both A and B.

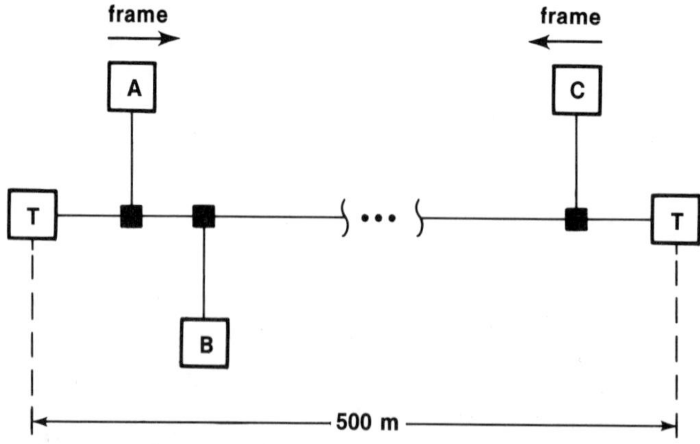

FIGURE 13–7   Collisions on a LAN

Both A and C have a frame destined for station B. Signals move on the coaxial cable at approximately 4 ns/m. If C starts transmitting first, the time for the leading edge of C's frame to reach B is

$$500 \text{ m} * 4 \text{ ns/m} = 2000 \text{ ns}$$

At this point there is still an infinitesimal time left before the frame reaches A. Suppose A begins transmitting at exactly this instant. We have a collision, and the signal received by B will be garbled. Station A will know about it very shortly, but how long will it be before C is aware of it? C cannot detect A's frame for another 2000 ns. Also, if C stops transmitting before the signal from A reaches it, C will not notice the collision. (The precise definition of a "detected collision" is this: a transmitting station noticing the presence of another carrier on the cable simultaneously with its own.) C's frame must therefore last for at least $2000 + 2000 = 4000$ ns. That is, twice the end-to-end propagation time.

For actual networks the end-to-end time calculation must account for delays in transceivers, transceiver cables, and repeaters, as well as propagation down the coaxial cable. For Ethernet the longest possible two-way delay turns out to be approximately 45 μs. The minimum frame time is

$$64 \text{ bytes} * 8 \text{ bits/byte} * 0.1 \text{ } \mu\text{s/bit} = 51 \text{ } \mu\text{s}$$

which meets the requirement with a little bit to spare.

The CSMA/CD concept works well as long as there are not too many collisions. This is normally true for the short, intermittent bursts of data characteristic of applications involving terminals. If some stations on a network begin sending steady streams of long frames or if we attempt to put more and more terminals on the network, we will get to a point where collision frequency increases dramatically and operators experience long delays. In many cases users do not attempt to make the network carry more than 50% of theoretical capacity (10 Mbit/s) in order to avoid this congestion.

It is also interesting to note that, according to the five criteria listed at the beginning of Chapter 11, Ethernet is not a complete protocol. It provides framing and line control. It also provides error detection, but no recovery features. If a receiving data link layer detects a frame check sequence error, an indication is passed to the client layer, which is responsible for any recovery procedures. The issues of flow control and sequence control are also left to the client layer.

13·5 THE FILE SERVER, A LAN APPLICATION

Now that we know how to assemble a LAN, how can it be used to meet some of the needs listed at the beginning of this chapter? As an example, we will consider a network consisting of several personal computers (PCs) and one additional station called a *file server*. The network is illustrated in Figure 13–8.

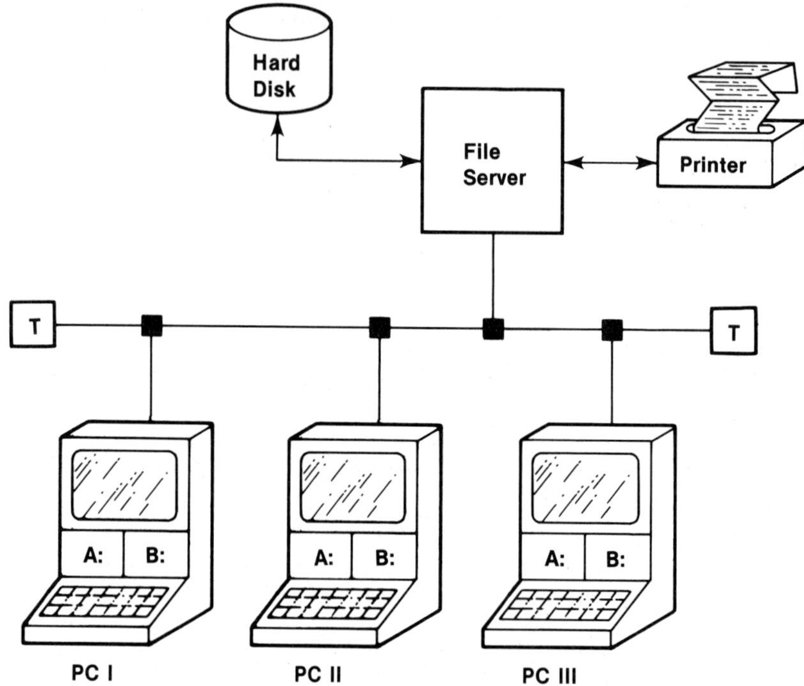

FIGURE 13–8   A LAN File Server System

The set of PCs can be a mixture of different models and hardware configurations, but, to simplify the example, we consider them to be identical. Each has
two diskette (floppy-disk) drives which are identified as A: and B:. They have
no other mass storage devices. The file server is another computer. Its data processing hardware may or may not be the same as the PCs, but there are two essential differences between the file server and the other stations:

File Server Software

The PCs run a single-user operating system which allows the user to run
one program at a time. The file server runs a time-shared system, similar to the
one described in Section 12·1, which allows concurrent execution of many programs.

File Server Peripherals

Rather than diskette, the file server's mass storage consists of "*hard disk*,"
usually of the variety known as "Winchester." There are two basic differences
between a Winchester disk and a floppy disk:

- *Storage.* A double-sided, doubled-density floppy disk holds approximately 350 000 bytes when formatted. The smallest Winchester disks hold 10 000 000 bytes, and their capacities go up to hundreds of megabytes.
- *Speed.* The time required to access data anywhere on a Winchester disk is typically less than 50 ms, and some models are significantly faster than that. Diskette drives may require hundreds of milliseconds to range over a much smaller volume of data.

The file server may also have high-speed, high-quality printer which is too expensive to allow it to be attached to a single-user system.

Typical network/file server systems provide PC users with features that would normally be associated with large host computers:

- shared access to high-performance disk drives
- printer spooling
- electronic mail

The operation of the file server is based on a *partitioning* of the hard disk and a special organization for the traffic on the network. In Figure 13–9 a hypothetical partitioning of a 10-Mbyte Winchester disk is illustrated. The traffic on this network flows in a restricted pattern. No direct PC to PC messages are allowed. All frames generated by the PCs have the file server as their destination address. The file server has the ability to send frames to any PC. Some additional network access software is also required in each user's PC.

Shared Disk Drive

Before the network was installed, drive C: did not exist on any of the PCs. The network access programs in the PCs recognize any request for data from a file on drive C: and translate it into a frame on the network. A program in the file server interprets this frame, finds the appropriate file on the Winchester disk and sends its contents back to the PC via the network. Since the file server must have the ability to deal with multiple, simultaneous requests of this type, the time-shared operating system is required. From the point of view of the user at the PC keyboard, drive C: appears to be local and the format of the com-

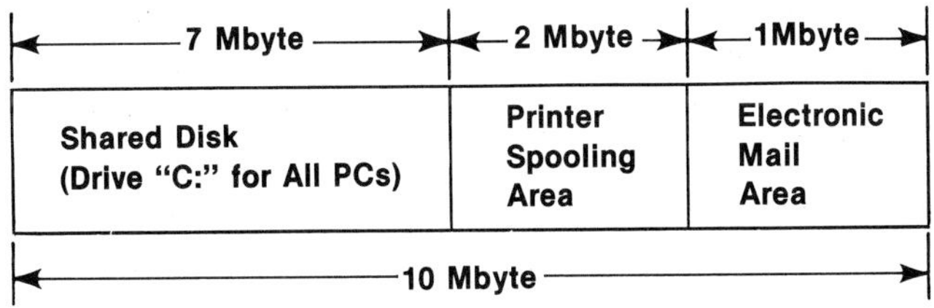

FIGURE 13–9 File Server Disk Partitioning

mands used for C: are no different from those used to access A: and B:. Because of the high speed of the network and the hard disk, response times are usually no worse than what would be expected from a diskette. In addition to greatly expanded disk storage, the shared drive offers several unique features:

- A program stored on the hard disk can be loaded, for execution, into the memory of any PC via the network. This is more convenient than distributing individual diskette copies to each user and, if the program must be modified, avoids the problem of redistribution.
- A shared data base may be stored on the hard disk, once again avoiding a distribution problem. There are possible problems here. Errors may occur if two or more users attempt to modify the same data item simultaneously. In multiuser data base programs this is usually avoided by allowing a user to "lock" a data item before attempting to modify it. The lock prevents other users from making changes until the first user unlocks the item. Typical PC data base management programs do not have this feature and therefore are not usable with shared file systems.

Printer Spooling

This feature allows all PC users to share the printer attached to the file server and, as an added bonus, allows them to use their PCs for other functions while the print jobs are being handled by the file server. Spooling operation is illustrated in Figure 13–10.

None of the PCs on the network have individual printers. When a user issues a print command, the network access programs in the PC intercept the command and send text that would normally go to a local printer to the file server instead. Several users may issue print commands simultaneously. If we allowed frames from users to be printed immediately, text lines from several jobs would be intermixed on the printer output. To prevent this we have the software configuration illustrated in Figure 13–10 (programs in the file server are shown as "clouds"). The PC user interface programs receive text frames for printing from the network. Print jobs are assembled frame by frame in the spooling area of the hard disk, but nothing is done with them until the entire job is received. At this point, the job is placed at the end of a waiting list (*spooler queue*) for the *printer driver* program. The printer driver does nothing until a job appears on its list. It then sends text to the printer until the list is empty. As jobs are printed, their text is discarded from the disk.

In addition to sharing the cost of a printer among several users, this system has other advantages. In a typical single-user system, nothing else can go on while a print job is running. Since PC printers are usually rather slow, printing may involve a long period during which the operator cannot use his PC for anything else. With the spooling system, text is transferred to the hard disk as fast as the network can accept it. For large print jobs, the network transfer time will

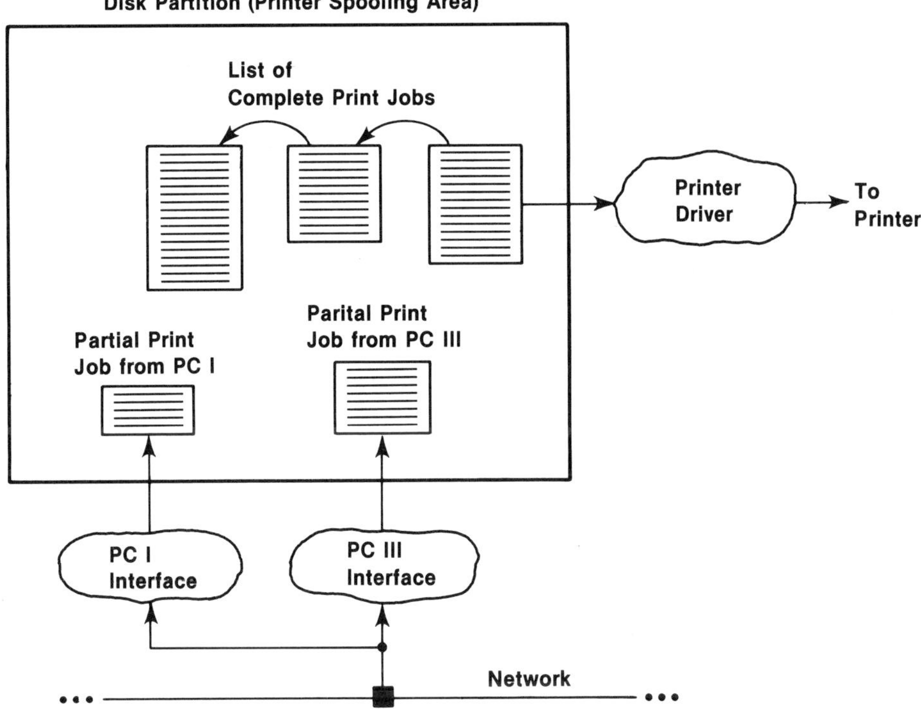

FIGURE 13–10   Printer Spooling

be much less than the print time. Once the job is in the spooler queue, the PC user is free to move on to some other program.

Electronic Mail

Electronic mail (see Figure 13–11) operates in a manner similar to the spooling mechanism. Each user has a queue of messages waiting to be read. (For privacy, mail messages are addressed to a "user id" rather than a "station id.") Any user can generate a message which is sent via the network to the *message builder* in the file server. The message builder places complete messages on the queue to the appropriate user.

Nothing is delivered until a user enters a request to read his mail. Messages are then sent to the addressee via the network and removed from the electronic mail area of the disk.

One further note: in spite of the way Figures 13–10 and 13–11 are drawn, programs such as the PC spooler interfaces, message builder, and message de-

Disk Partition (Electronic Mail Area)

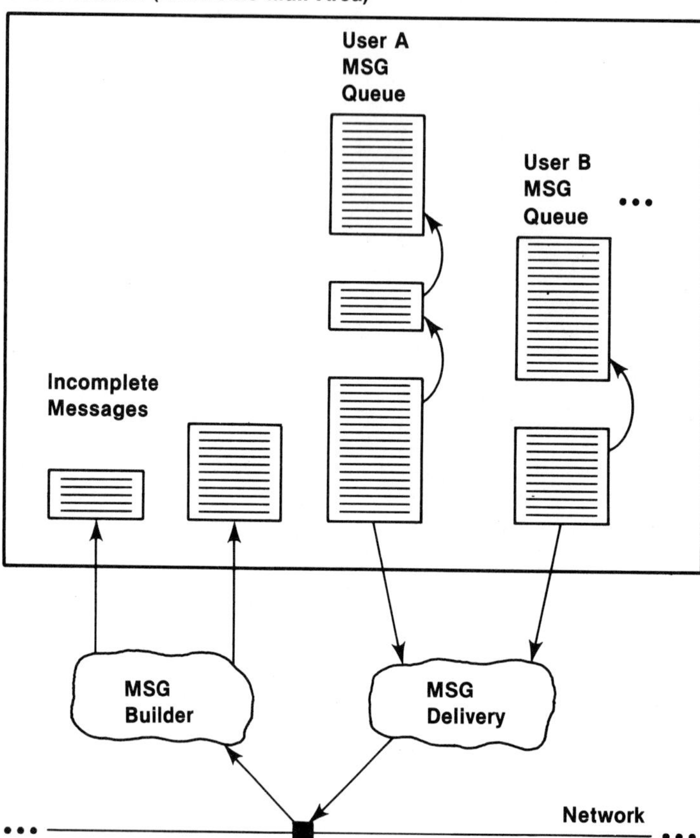

FIGURE 13–11   Electronic Mail

livery are part of the Ethernet client layer and do not directly communicate with the coaxial cable.

It is interesting to contrast the network/file-server concept (Figure 13–8) with the older multiuser host-computer concept (Figure 12–1). The network/file-server system is much more flexible. For a given application a large host computer has the capacity to handle a particular number of users. If fewer users than this number are attached, the host computer is underutilized; if more are attached, users experience delays. In the network/file server, each user continues to run programs in his own PC. Therefore, as we add more users, we add more processing power. (This simple generalization is, unfortunately, not the full story. In both the host computer and the network/file-server, the single access point to the disk remains a traffic "bottleneck." In applications which use the disk heavily, adding computational power may not yield any improvement in performance.) On the other hand, if memory is limited in the PCs, it may be impossible to run some programs on the file server configuration that could be handled by the host computer.

13·6 THE TOKEN RING, ANOTHER APPROACH

A totally different design concept for a LAN is the token ring. In this case the stations are arranged in a ring (see Figure 13–12). The stations are not merely taps on a cable and data flows in only one direction around the ring. Each station receives frames from the previous station and retransmits them to the next station. This is necessary because, in some cases, the protocol requires a station to transmit something other than what it receives.

The line control protocol operates as follows:

- There is a special bit pattern called the *token* which circulates around the ring. If a station has nothing to send, it will merely retransmit the token when it receives it.
- If a station has a frame to send, it waits until the next time it receives the token and then removes it from the network. The station then transmits the frame to the next station. As in Ethernet, frames have a source and destination address. Each station passes the frame to the next station around the ring. All stations other than the one specified in the destination address ignore the contents of the frame.
- When the originating station receives its own frame, it knows that the message has made a complete cycle and has therefore been "broadcast" to all other stations. The originating station now removes the frame from the network and reinserts the token.

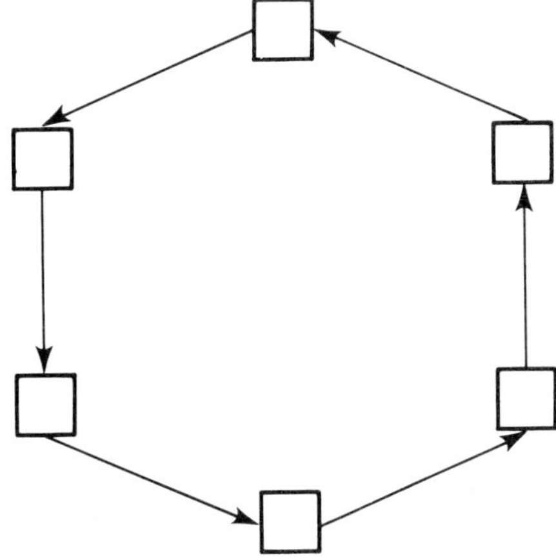

FIGURE 13–12   A Ring LAN

The primary advantage of this concept is that, without the use of a master station, we regain the precise control over who is allowed to transmit that was present in a polling situation. Under heavy traffic conditions the token ring will perform better than a free-for-all like Ethernet. (Better performance means that, for high traffic levels, data transfer rate on the ring will be closer to the theoretical maximum capacity of the medium than it would be for Ethernet.)

We have, however, paid for this performances with increased complexity, and reliability becomes a more serious consideration. If an error changes the token pattern, it stops circulating, and traffic on the ring comes to a complete stop. Also, unless a failed station is bypassed, no data traffic is possible. Solutions to these problems have been implemented, but, as of the present, ring networks have not gained the popularity of "bus" networks such as Ethernet.

REVIEW QUESTIONS

1. Draw the Manchester encoded waveform for the following bit sequences:
   a. 1 0 1 0 1 0 1 1
   b. 0 0 0 0 1 1 1 1
   c. 0 1 1 1 0 1 0 0

2. Describe some of the problems that might be associated with a LAN using airborne radio frequency signals.

3. If Ethernet stations were installed with minimum separation, how many could be attached to a single segment of coaxial cable?

4. Ethernet stations A and C are separated by 500 m of coaxial cable. Station B is exactly at the midpoint of the segment. A and C both have frames addressed to B and begin transmitting at the same instant.
   a. How much time is required for these signals to reach B?
   b. Will B receive any one of these frames without interference from the other?
   c. How much time (after the start of transmission) is required for A and C to detect the presence of the collision?
   d. For this particular configuration, what is the minimum frame length?

5. How many double-sided, double-density floppy disks would be required to store all the data on a 20-Mbyte Winchester disk?

6. A PC has a 30-char/s local printer.
   a. How much time is required to print a 6000-character text file on the local printer?
   b. This same file is transferred, via LAN, to the spooling partition of a file server disk. If each LAN frame contained 100 characters of text in the user data field and it takes an average of 50 ms to write each frame on the disk, how much time is consumed by the network transfer?
   c. How much sooner is the PC available for new work when the spooler is used?

Appendix A:

DIGITAL LOGIC CIRCUITS AND TTL

Bipolar transistor-transistor logic (TTL) is a circuit design and manufacturing technique for small- and medium-scale digital integrated circuits. Within this family of devices there are several subfamilies which provide a choice of operating speed and power consumption level, but, for all members of the TTL family, there is a standardized definition of the voltage levels representing the two possible states of a digital circuit:

> 1: approximately +4 VDC
> 0: approximately +0.5 VDC

Other conventions are possible, and there are logic circuit families based on different manufacturing techniques, but the previous definitions are, by far, the most common for general purpose digital designs. Throughout this book we will only deal with 1s and 0s, and we therefore do not have to be especially concerned with exact voltage levels and circuit details. From this point of view there are two basic component types that we have to understand: *gates* and *flip-flops*.

GATES

Gates are digital circuits with one output and one or more inputs. The operation of a gate is defined by its *truth table*, which is a list that specifies the state of the output (1 or 0) for every possible combination of input states. There are three basic varieties of gates.

AND Gates

The truth table and schematic diagram symbol for a two-input AND gate is given in Figure A-1.

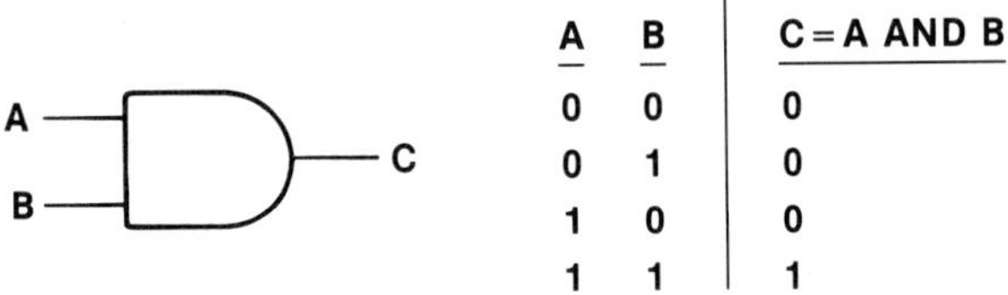

A	B	C = A AND B
0	0	0
0	1	0
1	0	0
1	1	1

FIGURE A–1   The AND Gate

The essential feature is that the output is at 1 only if all of the inputs are simultaneously at 1. In the algebra of digital logic circuits this is expressed as

$$A \quad AND \quad B = C$$

This concept is easily extended to AND gates with three or more inputs.

OR Gates

The truth table and schematic diagram symbol for a two-input OR gate is given in Figure A-2.

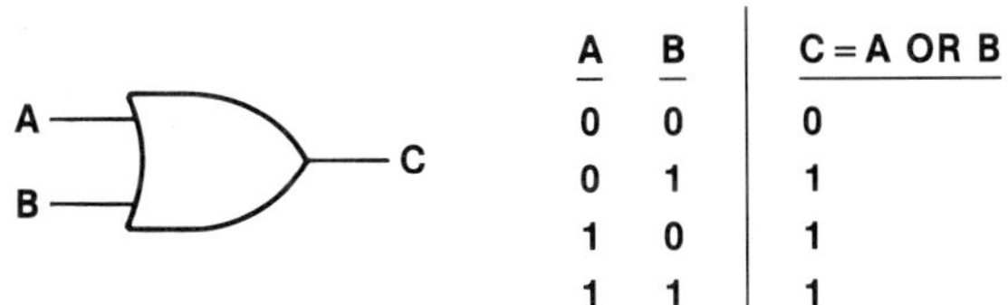

A	B	C = A OR B
0	0	0
0	1	1
1	0	1
1	1	1

FIGURE A–2   The OR Gate

In this case the output is at 1 if any one or more of the inputs are at 1. Algebraically, this is expressed as

$$A \quad OR \quad B = C$$

Once again, this concept is easily extended to gates with more than two inputs.

Inverters

The inverter is a one-input gate whose truth table and schematic symbol are shown in Figure A-3.

A	$B = \bar{A}$
0	1
1	0

FIGURE A–3   The Inverter

The output is always in a state opposite that of the input. Algebraically, the term NOT expresses this state inversion (where NOT A is represented by $\bar{A}$)

$$\bar{A} = B$$

Inverters always have only one input.

Given these three basic gate types, we can synthesize more complex functions. For example, the logic equation

$$A \ \ AND \ \ \bar{B} = C$$

could be implemented as shown in Figure A-4.

A	B	$C = A \ AND \ \bar{B}$
0	0	0
0	1	0
1	0	1
1	1	0

FIGURE A–4   More Complex Functions

In this case the output is at 1 only if input A is at 1 and input B is simultaneously at 0. Functions like this occur frequently in the design of digital equipment, and, for purposes of illustration, they are often depicted as a single schematic symbol by using a "bubble" at a gate's input or output to represent the presence of an inversion. For this example we have Figure A-5.

FIGURE A–5   A Single Complex Gate

Another example is Figure A-6.

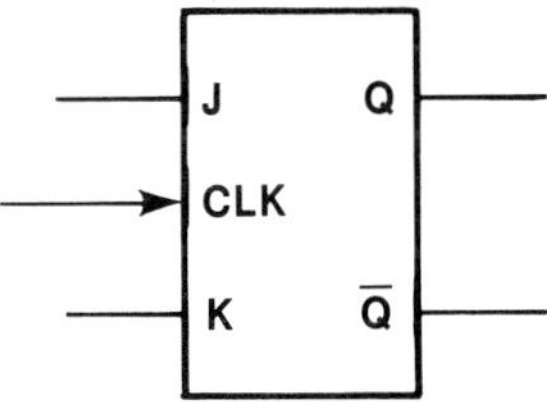

A	B	C = $\overline{\text{(A OR B)}}$
0	0	1
0	1	0
1	0	0
1	1	0

FIGURE A–6   Another Complex Function

Some of these complex functions may be available as actual integrated circuit components and others may not. In the latter case they may always be synthesized from the three basic types.

FLIP-FLOPS

The essential feature of a flip-flop is its "data storage" capability. While the output of a gate changes at the instant that its inputs change, the output of a flip-flop may represent the condition of its inputs at some time in the past.

There are several varieties of flip-flops. The one we describe is the most versatile: the JK flip-flop. This device has three inputs (J, K, and CLOCK) and two outputs (Q and $\overline{Q}$). It is schematically represented in Figure A-7.

FIGURE A–7   The JK Flip-Flop

The JK flip-flop operates as follows:

- The binary state of $\overline{Q}$ is always opposite the state of Q.
- The states of the outputs will not change until a pulse appears on the CLOCK input. The state change will actually take place on the 1 to 0 transition of the clock pulse.

• At the time of the 1 to 0 transition of the CLOCK input, the outputs behave as shown in Figure A-8.

Inputs	Output Behavior
J = K = 0	No Change
J = K = 1	Q and $\overline{Q}$ both assume states opposite their previous values
J ≠ K	Q assumes the state of the J input $\overline{Q}$ assumes the state of the K input

FIGURE A–8   Operation of a JK Flip-Flop

Many useful devices may be constructed from flip-flops. One of the most common (and one which appears frequently in this book) is the shift register shown in Figure A-9.

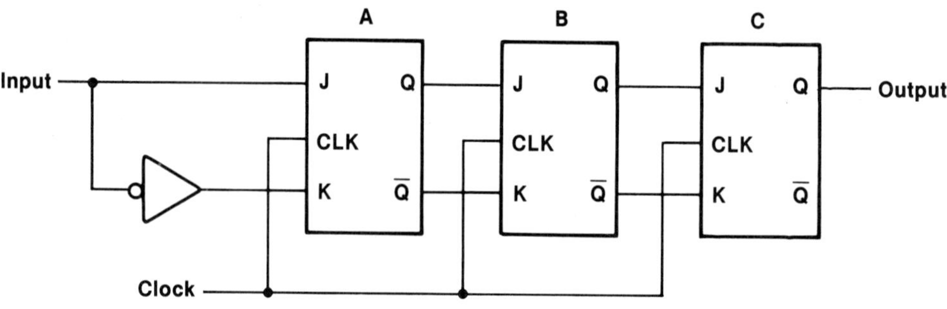

FIGURE A–9   The Shift Register

Each flip-flop in the shift register is called at *stage*. This circuit configuration assures that the state of J will always be different from that of K for all stages. Therefore, following each clock pulse, the outputs of each stage assume the values that were present at that stage's inputs before the clock pulse was applied. The state of INPUT therefore appears to "shift" from left to right. For example, the Q output of stage C represents the state of INPUT three clock pulses earlier. Stage B represents the state of INPUT two clock pulses earlier, and so on.

Typical TTL JK flip-flops also allow the stages of a shift register to be loaded simultaneously (in parallel) from some external device (usually a microprocessor). When clock pulses are applied following the load operation, the shift register becomes a "parallel-to-serial converter."

Appendix B:

THE BINARY AND HEXADECIMAL NUMBER SYSTEMS

The number system we use for everyday calculation is called the *decimal* system because it is built around the number 10. Let us look at a typical 4-digit decimal number and see how it is constructed:

$$5478$$
$$= \quad 5000 + 400 + 70 + 8$$
$$= \quad 5 * 1000 + 4 * 100 + 7 * 10 + 8 * 1$$
$$= \quad 5 * 10^3 + 4 * 10^2 + 7 * 10^1 + 8 * 10^0$$

Two concepts are at work here:

1. *Digit Value*. Each digit is selected from a set of 10 possible values (0–9).
2. *Positional Weighting*. Each digit is multiplied by a weighting factor (10^0, 10^1, 10^2, …), whose value increases as we move toward the left.

The number 10 (the *base* of the decimal system) comes into play in two places:

- There are 10 possible choices (0, 1, …, 9) for the value of each digit in the decimal number.
- As we move to the left, each weighting factor is a successively higher "power of 10."

We can generalize this concept and speak about numbers constructed around an arbitrarily selected base which we call B:

- Digits may be selected from a set of B different values ranging from 0 to B−1.

- As we move to the left in a base B number, the weighting factors are successively higher powers of B.

A general formula for a 4-digit number in any base is therefore

$$D\_3 * B^3 + D\_2 * B^2 + D\_1 * B^1 + D\_0 * B^0$$

where D_3, D_2, D_1, and D_0 are the digits of the number and must be selected from the set 0, 1, …, B − 1. These concepts, of course, may be extended to numbers larger or smaller than four digits.

An example of a simple and useful nondecimal number system is the *binary* system, whose base (B) is 2. From the previous material we can draw the following conclusions about the binary system:

- The only possible digit values are 0 and 1.
- The weighting factors are

$$2^0 = 1$$
$$2^1 = 2$$
$$2^2 = 4$$
$$2^3 = 8$$
$$2^4 = 16$$
$$\vdots$$

Let us take a typical binary number and find its decimal equivalent:

$$1\,0\,0\,1\,1$$
$$= 1 * 2^4 + 0 * 2^3 + 0 * 2^2 + 1 * 2^1 + 1 * 2^0$$
$$= 1 * 16 + 0 * 8 + 0 * 4 + 1 * 2 + 1 * 1$$
$$= 16 + 2 + 1$$
$$= 19$$

Table B–1 lists the first thirty-two binary numbers and their decimal equivalents. The student should examine this table in detail, paying particular attention to the counting sequence. Notice how groups of 1s "roll over" to a single 1 in the next higher digit position followed by a series of 0s. The student should also verify the values of the decimal equivalents by the preceding method.

The binary number system is used universally in digital computing and communications equipment. In fact, the familiar term *bit* is a contraction of *binary digit*. Most of this equipment handles data in a 8-bit groups called

TABLE B–1

BINARY	DECIMAL	BINARY	DECIMAL
00000	0	10000	16
00001	1	10001	17
00010	2	10010	18
00011	3	10011	19
00100	4	10100	20
00101	5	10101	21
00110	6	10110	22
00111	7	10111	23
01000	8	11000	24
01001	9	11001	25
01010	10	11010	26
01011	11	11011	27
01100	12	11100	28
01101	13	11101	29
01110	14	11110	30
01111	15	11111	31

bytes. Bytes are handy data units, but they create a "communications problem" for people. For example, read the 3 bytes in Table B-2 to a friend (who is not looking at this book) and ask the friend to accurately repeat them:

TABLE B-2

1	0	0	0	1	0	1	1
0	1	1	0	1	0	1	0
1	1	1	1	0	0	1	0

During the design and testing of digital equipment, members of the project group must frequently communicate such strings of information to other project members. What is required is a more concise and mnemonic way of expressing binary numbers.

Suppose we break each byte in Table B-2 into two groups of four bits. From the left side of Table B-1 we can see that any possible 4-bit binary number can be represented by a value in the range 0, 1, ..., 15. This suggests that each 4-bit group may be considered as a single digit in a base 16 (*hexadecimal*) number system. The familiar number systems do not have any one-character symbols for digit values greater than 9, so we must create them. The choice, which is a standard throughout the computer industry, is the use of A, B, ..., F to represent the digit values 10, 11, ..., 15. Table B–3 presents the full list of digits in the hexadecimal system.

TABLE B-3

HEXADECIMAL DIGIT	BINARY EQUIVALENT
0	0000
1	0001
2	0010
3	0011
4	0100
5	0101
6	0110
7	0111
8	1000
9	1001
A	1010
B	1011
C	1100
D	1101
E	1110
F	1111

All bytes can now be represented as a combination of two hexadecimal digits, and we can concisely express the data in Table B-2 as

8B
6A
F2

Index